MAKING CHANGE IN MATHEMATICS EDUCATION

Learning from the Field

edited by

Joan Ferrini-Mundy
University of New Hampshire

Karen Graham
University of New Hampshire

Loren Johnson
University of California at Santa Barbara

Geoffrey Mills
Southern Oregon University, Ashland, C

National Council of Teachers of Mathematics
Reston, Virginia

Library of Congress Cataloging-in-Publication Data:

Making change in mathematics education : learning from the field /
edited by Joan Ferrini-Mundy ... [et al.].
p. cm.
Includes bibliographical references.
ISBN 0-87353-442-5
1. Mathematics—Study and teaching—United States.
I. Ferrini-Mundy, Joan, 1954–
QA13.N143 1998
510′.71′073—dc21

98-46363
CIP

The publications of the National Council of Teachers of Mathematics present a variety of viewpoints. The views expressed or implied in this publication, unless otherwise noted, should not be interpreted as official positions of the Council.

Printed in the United States of America

TABLE OF CONTENTS

PREFACE

With its publication of the *Curriculum and Evaluation Standards for School Mathematics* in 1989, the National Council of Teachers of Mathematics (NCTM) launched the standards movement. The *Curriculum Standards* represents the best consensus effort of the field to describe a vision for grades K–12 mathematics teaching and learning. In practice, the process of moving toward that vision has been both challenging and rewarding for teachers, for curriculum developers, for community members, for district administrators, and for policymakers. This volume is intended as a tool for those practitioners—a source of ideas and a common site for the discussion of the deep and difficult issues involved in making the vision of the *Curriculum Standards* real in classrooms.

The Recognizing and Recording Reform in Mathematics Education project (R^3M) was initiated in 1992 by NCTM, with funding from the Exxon Education Foundation, as an effort to document and gain insight into the process of making Standards-based change in mathematics teaching and learning. Through the qualitative study of seventeen sites around the country, the R^3M documenter team assembled site-based stories addressing mathematical vision, pedagogical vision, contextual features, effects on students, the evolution of reform efforts, and the role of standards. The teachers, administrators, parents, and students in these sites cooperated generously with their time and openness in describing the road to reform—a road that is sometimes rocky and winding, sometimes smooth and straight. The insights they shared with us, together with the interpretation provided by the R^3M team, form the basis for this book.

The R^3M project was designed and conducted as a qualitative research study. The research report from this project is available in the *Journal for Research in Mathematics Education* Monograph Number 8, *Recognizing and Recording Reform in Mathematics Education: Issues and Implications* (Ferrini-Mundy and Schram 1997). A brief description of methodology is appropriate here. Following a national call for nominations of school or district sites that were attempting to implement the *Curriculum Standards,* the R^3M advisory committee and staff selected sites to be visited in fall 1992. A team of more than twenty documenters—including mathematics education researchers, teacher educators, educational anthropologists, school administrators, teachers, and graduate students—was assembled to design the project methodology and perspective. Beginning in fall 1992, we conducted visits, lasting from two to four days, to each of the seventeen sites. Two documenters visited each site and interviewed teachers, students, administrators, and community members about the changes and practices in mathematics education in the sites. We were especially interested in the role of standards in the sites. Following the visits, the documenters wrote scenarios about the sites. These scenarios were relatively short descriptions of some interesting features of a site relevant to its efforts in mathematics education. The scenarios were not fully interpreted but rather offered a view from the site's perspective by drawing heavily on actual quotations from interviews and observations from field notes. For some sites, full case studies were written. At all sites, individuals have been given pseudonyms, and we have done our best to protect the confidentiality and anonymity of all who participated in the study. Excerpts from the scenarios and the case studies form the basis for much of this volume.

A goal of the R^{3}M project from the outset was to give practitioners—K–12 teachers and administrators—useful images about mathematics education reform and helpful accounts from sites that have been engaged in sustained efforts to improve mathematics education. The R^{3}M database includes a wealth of such images and accounts, and we use them in addressing central questions that are currently at the heart of mathematics education improvement efforts. Each chapter addresses important questions that surface for practitioners seeking to change and improve mathematics education and draws heavily on material from the R^{3}M database, case studies, and scenarios to present rich illustrations of the issues.

For the R^{3}M data gathering, the question "What does it mean to implement the *Standards*?" was fundamental. We can use our sites' experiences to discuss what implementation has come to mean at the state framework level, at the district guide level, and in the classroom. In chapter 1, Johnson and Mills highlight the dilemmas that we and others faced even in deciding whether a lesson is "Standards based." The R^{3}M sites generally addressed pedagogical issues more directly and extensively than they did mathematical ones. The kinds of tensions that emerged around how the "basic skills" should be balanced, how teachers were interpreting problem solving and communication, and how certain Standards-based mathematics topics were introduced first as curricular add-ons are examined in chapter 2 by Graham and Ferrucci. The R^{3}M sites drew on a variety of approaches to professional development, funding, and choosing curriculum materials to initiate and sustain their efforts. These different reform structures might be characterized as ranging from static to dynamic. In chapter 3, Tinto and Masingila illustrate these approaches and discuss the implications and trade-offs involved in each situation.

In all the schools and districts in the R^{3}M study, identifiable leadership for the mathematics education reform agenda was a crucial factor in their success. These leaders ranged from self-appointed individuals to the superintendent. The various approaches and how they worked are taken up by Masingila and Tinto in chapter 4. Most of the R^{3}M sites were far enough along in their reform efforts when we visited that they had grappled directly with the challenges of informing and involving the community and of garnering its support. This happened in various ways with various outcomes and is discussed by Graham and Johnson in chapter 5.

As Lambdin describes in chapter 6, some of the R^3M sites used assessment as a lever for change in mathematics teaching practice. We discuss the challenges and positive features of such an approach, drawing heavily on site examples. Several of the R^3M sites entered their reform efforts with a primary focus on raising expectations and providing rich mathematical experiences for *all* students, especially groups not always well served by mathematics programs. Some bold efforts were made to change curriculum, to introduce algebra for all, and to eliminate general-mathematics tracks in our sites; Johnson considers these efforts in chapter 7.

In chapter 8, Koch reflects on the many challenges, tensions, and obstacles to making significant and deep Standards-like change in mathematics classrooms and programs.

In conclusion, Ferrini-Mundy summarizes the lessons learned from the R^3M sites, especially getting started in Standards-based reform, what to expect along the way, and how to keep it going. In addition, the R^3M project has taught us a great deal about how the documents are interpreted in the field, and this information is summarized as well. Finally, we include some comments about the current status of standards-based reform generally and the possible role of mathematics in that context.

Debates about the appropriateness and direction of mathematics education reform, and about standards, are lively and rich. Our hope is that through examples from the sites, we can illuminate some of the troubling issues currently under discussion. We have learned a great deal in this project; most notably, we have learned that change is slow and difficult, needs sustained attention and widespread engagement, and can make a positive difference for students of mathematics at all levels.

We encourage readers to use this book in various ways. Certainly we hope that it will appeal to a wide audience interested in an inside look at mathematics education in real schools and districts. We learned from our sites that in many instances, change begins because teachers, as individuals and in small groups, develop an interest in standards and reform ideas and find ways to explore them, individually and with colleagues. Very little literature is available for teachers and administrators to use in these informal collaborative settings to promote a deeper discussion of the issues that a school or district will face as it gears up to focus on mathematics education. In some sense, this book has been conceptualized as "cases" to promote discussion about issues in reform—an analogue of sorts to "cases" for teachers to use to consider specific teaching ideas. We hope that you will find this a rich and provocative set of materials that pushes your own thinking and that of your colleagues.

Joan Ferrini-Mundy
Karen Graham
Loren Johnson
Geoffrey Mills

ACKNOWLEDGMENTS

We gratefully acknowledge the generous funding of the Exxon Education Foundation and the support and confidence supplied by Mike Dooley and Bob Witte from the earliest stages of this project. In addition, NCTM staffers Marilyn Hala and Laura Dove have been invaluable in their logistical assistance. We owe special thanks to the R^{3}M Advisory Board members, led by Mary Lindquist, for the wisdom in thinking through the very challenging design and methodology issues for this study. The team of documenters worked heroically to build a collaborative project perspective and to carry out a massive data-collection effort in a relatively short time. The project reports, written in the form of "scenarios," serve as a basis for the chapters presented here. The reports were originally drafted by one or more of the project documenters (see page 146). The portions of the scenarios that appear in the chapters are the product of group discussions, revisions, and editing by the authors of the chapters. Thus, we have not attributed any of the scenario material to particular authors. Some of the scenarios are used in the *Journal for Research in Mathematics Education* monograph (Ferrini-Mundy and Schram 1997), and excerpts from that publication appear here as well. A special thanks goes to Loren Johnson for his unfailing loyalty to this project from start to finish. And finally, without the candor and generosity of the teachers, administrators, students, and community members in the R^{3}M sites, we would have no insights to share.

The views expressed in these chapters are those of the authors and do not necessarily represent the positions of the National Council of Teachers of Mathematics or the Exxon Education Foundation. Although I am currently working at the National Research Council, my involvement in the R^{3}M project is as a member of the faculty at the University of New Hampshire; therefore, the views presented here are not those of the National Research Council.

Joan Ferrini-Mundy
Project Director

DESCRIPTION OF THE R³M SITES

(Sites marked with an asterisk were selected for second visits and are discussed in detailed case studies in Ferrini-Mundy and Schram [1997]. Some material from those case studies appears here. Descriptions reflect sites at time of visits.)

Bedford Middle School

Bedford Middle School is situated in one of the poorest areas of a large urban district. The majority of the school's population is Hispanic. (Caucasians compose less than 10 percent of the districtwide population.) Ninety percent of the district's nearly 14 000 students qualify for free lunches. There are three other middle schools in this district, along with fourteen elementary schools and two high schools. Additionally, the district has a prekindergarten program, two alternative secondary schools, and a center for students with multiple handicaps.

Billabong Elementary School

Billabong Elementary School is a large grades K–8 school that has embraced the use of technology as an important component of its mathematics curriculum reform efforts. In large part, this technology has been made possible through school-business partnerships that have been forged by the school's principal. In many ways, the school looks different from traditional schools. Visitors are invited to look into classrooms through large windows. Kindergarten through third-grade classrooms characteristically have six Macintosh computers, scanners, and color printers and are networked with the school's library, which gives them access to the extensive CD-ROM collection. The fourth-through-eighth-grade classrooms have an additional six computers. In one class all the children are given an individual Powerbook to use for the year.

Chelmsville and Prairie Elementary Schools

Chelmsville and Prairie are located in a rural agricultural and lumbering area that is only nine miles from a state university. Students are located in three schools split by grades K, 1–3, and 4–6. The total student population is approximately 650 students. The school population consists of only Caucasians. Because the state has no waiting period for unemployment benefits, the number of people moving to Chelmsville has sharply increased. The income level is low to middle.

Deep Brook Elementary School*

Deep Brook is part of a district that is classified rural but is actually closer to suburban. Most of the families are middle to upper-middle class. The local nuclear power plant has been taken over by the state, and the district is now faced with declining local revenues. More than 90 percent of the students who graduate from high school go on to higher education. The minority population at Deep Brook is less than 1 percent and is composed of Asian and African American children.

Desert View High School*

Desert View High is a grades 9–12 school of approximately 2 200 students, with a 50-50 mix of Hispanic and Caucasian students. A number of the non-English-speaking students study mathematics in English-as-a-second-language classes. This area is primarily agricultural, but a large federal military installation is part of the school's attendance area. A state university, also located in Desert View, is a short distance from the high school's campus.

East Collins High School*

East Collins High School is located just an hour from a major southern city in a rural setting where homes are replacing farms in a rapidly growing area. The school, just five years old, is already filled to its limits. The 1 125 students of the school come from different cultures and different socioeconomic levels. Approximately 30 percent of the student population is African American. A few Asians also attend the high school. The remainder of the student body is Caucasian.

Garnett School

The Garnett School is located in an old historic area of a large East Coast city. Most of the students who attend the school live in the subsidized housing project that surrounds the school. For many students, the school is a safe haven from the drugs, street crime, and vandalism that fill their lives. The Garnett faculty teaches the children in hopes that they will be able to better themselves and move away from the project, but the teachers believe that the environment has already taken its toll on many of the children, particularly the drug-addicted babies who are now starting to fill seats in the elementary schools.

Golden School

Golden is a magnet school with a racial balance equivalent to that of the middle-class area in which it is situated (approximately 45 percent Caucasian, 25 percent Hispanic [non–Puerto Rican], 20 percent African American, 10 percent Asian). Immigrants from eighty-seven countries live in the school's attendance area. Academically, the school population of approximately 1 350 students is 25 percent above average, 50 percent average, and 25 percent below average. The school serves students in grades 5 through 8, and classes are heterogeneous in composition. That 2 000 to 3 000 students compete every year for 400 open slots in this magnet school is an indication of its popularity among parents and students. The ninety teachers have an average class load of thirty students. Preservice undergraduates and teaching interns lessen that load considerably, so that small groups of students can be worked with on a regular basis.

Green Hills United School District

The Green Hills United School District has a grades K–12 student population of 11 126 students enrolled in twelve elementary, five middle, and two high schools. The minorities attending the district's schools account for approximately 5 percent of the total (Asian, African American, Hispanic, and Native American). During the site documentation, six schools (two of each type) were visited by the documenters. It is a suburban district in the Midwest with a mix of socioeconomic levels. Although most students are upper middle income, the southern part of the district has many students from farms and some of the poorest students in the county. The eighty-seven square miles that make up the school district encompass some of the fastest-growing suburbs in the nation, which contributes significantly to Green Hills's growth of almost 900 students a year. The district is supported by a school community that has very high expectations for its students and its schools.

Manzanita Middle School

Manzanita Middle School is part of a grades K–12 district that serves two cites in the suburbs of a large metropolitan area of the Northwest. There are 18 917 students attending school in the district, with schools and the student population distributed as follows: twenty-two elementary schools with 11 489 students, four middle schools with 2 591 students, and four high schools with 4 837 students. Minorities make up 15 percent of the total student population (Asian, 8.9 percent; Hispanic, 2.6 percent; African American, 2.4 percent; and Native American, 1.5 percent).

Oldburg Elementary School

The Oldburg Elementary School is located about an hour outside a major eastern metropolitan area in a middle-class community. Most students in the district come from single-family homes. The minority population of Asian, African American, and Hispanic students composes about 5 percent of the district's 3 500 students.

Parker Springs School District*

The Parker Springs School District has become one of the twenty largest districts in the nation and serves 110 230 students in 122 schools. There are 85 elementary schools, 21 middle schools, 12 high schools, and 4 alternative or exceptional schools. Forty-two percent of the district's students represent minority constituencies: 26 percent African American, 13 percent Hispanic, 3 percent Asian or Pacific Islander, and 0.5 percent Native American or Alaska Native.

Pinewood High School

In this magnet school the African American population is approximately 70 percent of the total student body. Pinewood High School is located in one of the more affluent areas of a large city and is part of a large county school district with many layers of administration. It is in the process of providing algebra and geometry for all its students.

Queensborough School District

Queensborough is a small rural district with 645 students in grades K–12. The two schools in the district are a grades K–6 elementary school and a grades 7–12 junior-senior high school. Queensborough is a vacation community, so the students are children of tradespeople and businesspeople who are supported by tourist activity. The education level of students' parents varies. The community is virtually all Caucasian.

Scottsville High School

This thirty-year-old high school is located in an upwardly mobile suburban area in the northern suburbs of a major midwestern metropolitan area. At the time of the R^3M visit, 92 percent of the student body was college bound, and the faculty at the school have always felt the pressure to maintain high standards. The school has experienced a recent influx of Russian and Polish children and has a large Korean population and a growing Hispanic population. Approximately 15 percent of the student population is Asian and 5 percent is Hispanic.

Southland Elementary School

Southland Elementary School (grades K–6) is located in a suburban community about twenty minutes from a major university. The socioeconomic level is average to above average, and fewer than 10 percent of the school's 700 students are from minority groups.

Timberkind Elementary School

This grades K–6 elementary school is located in the West Redwood School District, one of the largest school districts in the country. The population of the schools is two-thirds African American and one-third Caucasian, with a trace of Asian students. The school district has been under a continuing court order to desegregate for thirty-five years. Timberkind, once an all-Caucasian school, is located in a middle-class neighborhood and has a majority population of African American children who live in the inner city and are bused to the school.

Chapter 1

WHAT DOES IT MEAN TO IMPLEMENT THE NCTM *STANDARDS*?

Loren Johnson and Geoffrey Mills

Parallel *is a good word for it. In fact, I was doing a write up for somebody and a question was, "How do you use the* Standards *in your curriculum?" And when we first started Perfect Situation for Mathematics Learning [PSML], we had to prove to the parents that [PSML] was good, so the Standards were right there. And we were showing them this: "Look, we are doing what's new, and what's out there, and what's good for your students." And so we use the* Standards.

—Mathematics teacher
East Collins High School

When we designed or developed the mathematics program, we had the Standards that were discussed at times. We read the Standards, *became immersed in them, and probably, hopefully, the words that came out of our mouths to create our mathematics program were words that we had put in our minds from the* Standards *and from other good stuff, too, besides the* Standards.

—Curriculum coordinator
East Collins School District

After we really had a successful program with writing and mathematics, we went back to look at the Standards *and realized that it was aiming our students a lot more towards many of the Standards, and it was kind of neat. It was an affirmation that that's another reason why we must be doing the right thing. But it wasn't originally an attempt to address the Standards.*

—Mathematics department chair
Desert View High School

The format and organization of the Green Hills Curriculum Guide *was written with both the NCTM [National Council of Teachers of Mathematics]* Standards *and the state guidelines as models.*

—District mathematics coordinator
Green Hills School District

The district mathematics coordinator noted that the year of research and involving teachers and parents helped secure the approval [of the district's mathematics curriculum]. As part of this process, the teachers also approved the research committee's motion that the Curriculum and Evaluation Standards *of the National Council of Teachers of Mathematics and the state curriculum standards be adopted as the core component of the mathematics curriculum.*

—R^{3}M documenter

Just what does it mean to implement the NCTM's *Curriculum and Evaluation Standards for School Mathematics*?[1] All the foregoing quotations come from sites visited by documenters of the Recognizing and Recording Reform in Mathematics Education (R^{3}M) project. In each example, teachers and other central players in curricular development reported either that mathematics education in their schools was being driven by the *Curriculum and Evaluation Standards for School Mathematics* (NCTM 1989) or that the *Curriculum Standards* validated what they were doing in their mathematics programs. It became clear from the data gathered during the R^{3}M site visits that there were multiple interpretations of what people believed the *Curriculum Standards* to be and how those beliefs played out in practice. Perceptions held by people at the visited sites indicated that implementing the *Curriculum Standards* would mean different things to different people and that the local context in which change was occurring was instrumental in shaping those differences.

It was not uncommon for teachers, administrators, and policymakers to claim that they had adopted or implemented the *Standards* as their mathematics curriculum or to identify a particular text or program as representing *the* Standards. The comments introducing this chapter are but a small sample of those gathered during the fieldwork of the R^{3}M

[1] Throughout this book *Curriculum Standards* will be used to mean the *Curriculum and Evaluation Standards for School Mathematics* (NCTM 1989), *Professional Teaching Standards* will refer to the *Professional Standards for Teaching Mathematics* (NCTM 1991), and *Assessment Standards* will stand for *Assessment Standards for School Mathematics* (NCTM 1995). In addition, the following method of referring to Standards is used throughout the book. "*Standards*" (italic) stands for the books. "Standards" (roman, with an uppercase S) refers to the contents. Finally, "standards" refers to generic standards, in the sense of a general concept.

project. Frequently R^3M documenters found that people at the sites used the words *adopt* and *implement* interchangeably.

A broad-based effort to develop a Standards-based mathematics curriculum took planners in the Green Hills School District twenty-nine months. Yet in other districts, little advanced planning was done before mathematics programs, perceived locally as Standards based, were developed and implemented. Documenters observed something very different going on in each of the seventeen sites of the R^3M documentation study.

One might reasonably ask, What kinds of mathematics content and teaching practice did documenters expect to find during their field visits? Did any commonalties in implementation exist among the sites in spite of the observed differences? Each documenter held his or her preconceived notion of what Standards implementation should look like, and in spite of efforts to minimize these preconceptions, it would be difficult for documenters to resist making some judgment calls about a site's progress toward Standards implementation. It must be recognized that the R^3M documenters themselves came from varied backgrounds. What level of consensus might reasonably be expected with the mix of mathematics teachers, mathematics education researchers, and cultural anthropologists that made up this group of twenty-two documenters?

Before taking up these and other questions, we consider what the Standards and their intended purposes are.

WHAT ARE THE INTENDED PURPOSES OF THE STANDARDS?

The National Council of Teachers of Mathematics initiated a new era of content-based, professional organization–driven reform in mathematics education with the development and release of the draft of the *Curriculum and Evaluation Standards for School Mathematics* in 1987. The draft, mailed to 10 000 recipients, produced feedback of "more than 2 000 suggestions from parents, business leaders, teachers, and mathematicians," and the revised document was the product of extraordinary "review and consensus" (Romberg 1993). NCTM's *Curriculum Standards* (1989) was never intended to be a complete model of what one might expect to see in a reformed mathematics curriculum[2], although clearly, it included numerous vignettes and sample problems that gave glimpses

[2] According to the *Curriculum Standards* (NCTM 1989, p. 1), "A *curriculum* is an operational plan for instruction that details what mathematics students need to know, how students are to achieve the identified curricular goals, what teachers are to do to help students develop their mathematical knowledge, and the context in which learning and teaching occur."

of what content and pedagogy were supposed to look like. Rather, its authors saw the document as only "an initial step in the lengthy process of bringing about reform in school mathematics" (NCTM 1989, p. 7). They viewed the *Curriculum Standards* as "a vehicle that can serve as a basis for improving the teaching and learning of mathematics in America's schools" (p. 254) over the short term. In fact, the authors of the *Curriculum Standards* saw the initial version as short lived, with a life expectancy of only one decade before revision would be needed. The Standards 2000 Project, now under way, will produce this revision.

According to Apple (1992), the *Curriculum Standards* provides a "penumbra of vagueness." Is this vagueness essential for the *Curriculum Standards* to fulfill its mission in reforming mathematics teaching and learning? Given the vagueness and the various defensible interpretations of the *Curriculum Standards,* the process of documenting its impact on present mathematics content and teaching practices is complicated. Ferrini-Mundy and Johnson (1994, p. 6) address this concern:

> The *Standards* documents are clearly offered as a framework for discussion rather than a blueprint for change. They are intended to produce multiple interpretations and to invite discussion. At the same time, however, professionals in mathematics education certainly do hold strong opinions about what constitutes *Standards*-based mathematics teaching and learning. The matter of finding ways to judge, evaluate, and decide whether lessons, classrooms, teaching episodes, curricular materials, and school mathematics programs are truly "*Standards*-based" is fundamentally important, and necessary, yet also is paradoxical. The R^3M project stance has been that providing descriptions can contribute to these questions.

It is a complex matter to determine whether a mathematics curriculum is Standards-based. Professionals in mathematics education also have different ideas about how this congruence to the *Curriculum Standards* should play out in actual classroom practice. A request to describe what reform looks like will evoke an "it depends" reply because mathematics educators will select from a list of indicators those that they deem their favorites. For example, one might be concerned with the nature of discourse occurring in the classroom, whereas another might be on the lookout for the types of mathematical tasks being employed. They also hesitate to give a quick answer until they know more of the setting in which reform is taking place.

Johnson (1995) found that those involved in the reform of mathematics at the sites of the R^3M study perceived the *Curriculum Standards* in one of three ways: (1) as a validation of what they were doing in their mathematics program, (2) as a means for developing a different mathematics curriculum, or (3) as a tool for redirecting portions of their efforts in mathematics teaching. The role played by the *Curriculum Standards* in guiding reform activities depends on factors that often differ greatly from location to location. Looking at the physical manifestations of implementation may not be nearly as important as seeking answers to these questions: Does local implementation of the *Curriculum Standards* seek to achieve the main purposes[3] of the *Curriculum Standards*? Which ones?

DOES IMPLEMENTATION MEAN DIFFERENT THINGS AT DIFFERENT LEVELS?

The reform of mathematics teaching and learning takes place at different levels. The level of reform closest to where student learning occurs gives the best indication of how Standards-like a mathematics curriculum is. Typically, definite distinctions exist between changes taking place in the classroom and reform efforts taking place at the state and district levels. How does reform play out at each of these three levels: the state level, the district or school level, and the classroom level? How do the R^{3}M data give glimpses of what this reform might look like?

The Reform of Mathematics Education at the State Level

Generally, states have developed curricular frameworks and standards in grades K–12 mathematics, many of which are aligned with NCTM's *Curriculum Standards* (Joftus and Berman 1998). Currently, all states but one have adopted or are developing mathematics standards or frameworks for grades K–12 ("High Standards for All Children" 1998). What is included in these frameworks varies greatly from state to state. Several states have developed curricular materials that furnish detailed year-by-year mathematics content and benchmarks, whereas others provide only a brief guide of what content and pedagogy are important. Often, vignettes explaining pedagogical strategies are included (New Hampshire Department of Education 1995). Others, like the state of Indiana, have a document (Indiana Department of Education 1997) that teachers could use as an implementable curriculum because it details proficiency statements and indicators, essential skills, teaching and learning activities, and problem-solving strategies.

Those responsible for developing the Perfect Situation for Mathematics Learning (PSML) program at East Collins High School used

[3] The *Curriculum and Evaluation Standards* (NCTM 1989, p. 5) articulates "five general goals for all students: (1) that they learn to value mathematics, (2) that they become confident in their ability to do mathematics, (3) that they become mathematical problem solvers, (4) that they learn to communicate mathematically, and (5) that they learn to reason mathematically."

their state's recently adopted framework as a guide, as did the Green Hills School District. It is interesting to note that in both instances, the long-term curricular planning of mathematics content and instruction followed closely the guidelines adopted at the state levels.

It is important that teachers and policymakers understand the distinctions being made among the various levels of reform. Typically, states *adopt* a framework that is to be *adapted* by school districts, schools, and individual teachers. Thus the district or school mathematics curriculum would more closely resemble what might actually take place in classrooms.

Reform at the District Level

Before the *Curriculum Standards* was published, the county mathematics supervisor for Pinewood County found that her views on the reform of mathematics teaching and learning were unpopular with many of her colleagues. According to the supervisor, the *Curriculum Standards* gave credibility to her forward-looking views.

> *When the* Standards *came out in 1989, that summer, we sat down and said, "We need to do it differently." We already had the data that said we were not getting the job done. We knew that. The* Standards *gave us a vehicle by which we could depersonalize the change. Even now as I do in-services, I'm learning that whatever I say, black or white, handed out, this is from the* Standards. *Then people can't personalize it as my idea.*

The county mathematics supervisor had been in her position for a number of years. She had tried to eliminate what she saw as an elitist point of view toward mathematics and to face issues of equity squarely. It had been her goal to improve the quality of mathematics education for all students in the school district. She believed that low test scores in mathematics were too easily accepted and that teachers and others in the district conveyed a belief that the students of the county were just not able to do well in mathematics. Although her recommendations for change in content and pedagogy predated the *Curriculum Standards,* they were consistent with ideas found later in that document.

This county mathematics supervisor became the driving force for change in mathematics instruction in her district, yet she personally believed that whatever changes occurred in mathematics education at the district level should not appear to be the result of just one person's ideas. She saw the *Curriculum Standards* as lending credibility and guidance to the changes that were needed. She strongly believed that mathematics was an elitist subject that acted to distance underrepresented students from higher career goals.

There were other instances in which reform took root at the district level. In the Bedford School District, the focus of change was offering a core of algebra to all middle graders in the district, a change recommended in the *Curriculum Standards.* The Mathematics Teaching Project (MTP) was a districtwide effort aimed at ensuring that all students—particularly those from underprivileged or minority backgrounds—would be ready for a successful experience in algebra in grade 8 or 9. This goal was to be accomplished by showing elementary and middle school teachers how the concepts important to the study of algebra appear in other guises in mathematics throughout the first eight grades of school.

School and district administrators were very supportive of teachers during the time needed to develop and implement this core of algebra for all students. Although the school district is urban and poor, the MTP gave teachers materials and time for teaching, planning, and professional activity. These more tangible forms of support seemed to have been significant factors in the change process.

As a result of collaborative efforts between local university personnel and district-level administrators, funds from grants were obtained that afforded teachers opportunities to attend conferences. District and school administrators were able to establish school schedules that gave teachers the time to have a common planning period and to attend in-service meetings.

The district mathematics supervisor, one of the original organizers of the project, explained how the project evolved:

> *The first year we had fourteen teachers from each district. We took one middle school and its three main feeder schools. We tried to get as many teachers from the middle schools as we could, and I think in my district I had four volunteers from the middle school and ten from the elementary. I chose the middle school that was the lowest in my district, that we were already trying to help. I went there and explained, "Here is what we plan to do. Here is what we can offer."*

Can other poor school districts find some direction from the Bedford model? How important are the various levels of support? The MTP provided common planning time for teachers to collaborate and grow professionally. In what other ways might a school or district make such collaborative activity possible?

In the Green Hills School District, reforms centered on the development of a district mathematics curriculum. Teachers in the district saw that their broad base of participation on the various planning committees not only shaped the district mathematics curriculum but ensured community-wide acceptance of the Standards-based curriculum. Extensive preparation over twenty-nine months led to a new curriculum in their grades K–12 mathematics program that was shaped by teachers, administrators, parents, community people, business leaders, and students.

According to the district mathematics coordinator, this effort at changing the mathematics curriculum was very different from past attempts at curriculum revision. All stakeholders were involved in the process instead of just a few teachers, as in previous efforts. Four committees were formed—the mathematics curriculum research team, the mathematics curriculum writing committee, the mathematics text selection committee, and the mathematics technology scout team. These committees collaboratively carried out the initial stages of the curriculum development process that extended for more than two years before the actual program was implemented.

The school district spent an entire year just developing curriculum objectives for the grades K–12 mathematics curriculum and used the *Curriculum Standards* as the basis. The community input from various sectors first identified the mathematical needs of the community. People at the school believed that these needs fit perfectly with what the *Curriculum Standards* was saying, and the mathematics curriculum was developed by a broad base of representation and ratified by the teachers. District assessment was developed that would drive the instruction intended by the newly adopted curriculum objectives. Without the NCTM document, the mathematics supervisor believed that the finished product would not have turned out as complete as it did. The *Curriculum Standards* gave their efforts coherence.

Each teacher serving on the curriculum research committee, comprising forty-five members who represented all grades K–12 schools, was given a copy of the *Curriculum Standards* with the charge to read the document during the year of the committee's work. Exposure to the *Curriculum Standards* led to changes in how these teachers thought about mathematics instruction. The various curriculum committees had also decided that any viable mathematics curriculum would need to be ratified by all the teachers in the district, and this core of teachers, reporting regularly to their colleagues at the various schools, expedited the ratification process.

Without the *Curriculum Standards* to guide the efforts of the curriculum research committee, its members admitted that it would have been impossible to develop such a coherent document, one that all teachers later ratified. The supervisor said that the *Curriculum Standards* was the first thing that members of the committee looked at to prepare for their work. Teaching suggestions were later developed by elementary school teachers for their level. The *Curriculum Standards* also prompted the development of an assessment program that would drive the newly developed curriculum objectives of the district. Although a community-needs assessment preceded the study of the NCTM document, the Standards were used as criteria to check the value of those different needs.

A number of questions can be raised about such a model. Would the model for curriculum revision in Green Hills be suitable for other districts to follow? What are the costs of such an effort? What is involved in getting such large-scale participation across various sectors of the community? What mechanisms must be in place for such a massive undertaking to succeed? What role does needs assessment play in developing a mathematics curriculum that will be actively and willingly followed by teachers? How does one measure the "alignment" of the curriculum objectives with the *Curriculum Standards*?

We have discussed three examples in which a districtwide mathematics supervisor or coordinator exerted strong leadership to bring about a mathematics program that followed recommendations from the *Curriculum Standards*. The focus in the first instance was to erase the elitism that college-preparatory mathematics courses often create; the second was aimed at providing a core of algebra for students in grade 7 or 8; and the third was a broad-based effort to adapt districtwide mathematics curricula to the statewide framework.

School-Based Change

Data from the R^{3}M project do not support a one-model-suits-all plan for developing a Standards-based mathematics curriculum at the school or district level. As can be seen from the following illustrations, there are variations in who initiates the effort, in how the implementation of the curriculum proceeds, in what elements of standards are focal, and in how standards are used.

Reasons other than leadership at district levels precipitated Standards-based changes in mathematics. In one of our school-level sites, an elementary school principal became the visionary leader for changes in the mathematics program. Under his tutelage, many teachers at Deep Brook Elementary School also became leaders in working toward Standards-like reform in their school district and other districts through the mathematics workshops they conducted.

External funding was the impetus for teachers and administrators at East Collins High School to begin their PSML program. When local industry offered a fully equipped computer laboratory and mathematics software, teachers at East Collins High School believed that such a laboratory really did not epitomize what their ideal mathematics program should be. Influenced by their own teaching experiences and by what they had learned from reviewing the *Curriculum Standards,* they rejected the offer of support. "Then what do you want?" was the response from an industry spokesperson, which led to a partnership between the high school and industry.

East Collins teachers and administrators incorporated a planning and preparation phase that extended over approximately nine months. The planning team included high school mathematics teachers, an assistant principal, and a representative from local industry. With input from the school's mathematics teachers, PSML evolved as their curriculum model.

Consensus among the mathematics teachers was that the implementation of PSML should be gradual. During the first year, only three teachers were involved. By the third year, nearly every mathematics teacher had begun to implement the program.

In several districts, change to a Standards-like mathematics curriculum occurred first in individual classrooms. It was from this grassroots level of shifts in practice that schools first changed; later the impact was felt at the district level. How influential was studying the *Curriculum Standards* in these projects? Did the teachers' practices reflect some particular interpretation of the *Curriculum Standards*? Was the important feature of the innovation just the fact that the teachers came together and developed collegial, professional relationships? Did the rallying point really matter? How does personal ownership by teachers of an evolving program influence the program? Does such ownership strengthen the program?

WHAT DOES REFORM LOOK LIKE IN CLASSROOMS?

Teachers are continually reflecting on their practice, gathering data, interpreting the data, reinterpreting their practice, expanding their repertoire, and trying new approaches weighed against old "proven" practices that have withstood the test of time. This dynamic process is described as the "homeostasis of change" (Joyce, Hersh, and McKibben 1983): change is the norm, not the exception. Thus, on reflection, we think that the metaphor of a continuum is not an appropriate way of viewing the change process because it suggests that one eventually arrives at the end of the interpretation process. Our observations suggest that a spiral of change more accurately captures the interpretation and adaptation process involved with implementing the *Curriculum Standards*. Several examples from R^3M data illustrate the homeostasis of change that the documenters observed.

Ms. Wheeler was in her fourth year at Manzanita Middle School and described the evolution of her mathematics teaching philosophy as follows:

> *When I first started here four years ago, it was still traditional, using a book, doing some direct instruction, giving the children an assignment, they work on it, and so forth.... Two years ago I started a master's degree ... and it coincided with the mathematics reform effort in the district.*

Ms. Wheeler described her interpretation of the *Professional Teaching Standards* in terms of some tangible artifacts, including the use of worksheets. It is interesting to note that the mathematics teachers at this middle school did not use a textbook, which raised questions about what they did instead and the impact of this approach on the coherence of the mathematics curriculum in the school. The use of cooperative group work, the role of the teacher as a facilitator of learning, perseverance, the application of problem-solving techniques to real-life situations, the recognition that many problems have more than one correct answer, and writing about mathematics were found by Ms. Wheeler in the *Professional Teaching Standards.* It was with respect to this last characteristic that Ms. Wheeler expressed some concerns:

> *I'm trying to incorporate, which is one of the hardest things I'm struggling with, is how to get kids to write about what they're doing and become good writers. But you still get so many kids turning in "I multiplied and got the answer and that's just how you get it." So, I'm trying to get them to be more detailed about what they are writing, and they're getting better at it.*

It is easy to interpret the *Standards* documents as making certain recommendations, as did Ms. Wheeler in her emphasis on writing in mathematics classes. Teachers like Ms. Wheeler may criticize the *Standards* documents for not explaining in any depth *how* to do things. Where can teachers find the answers they seek?

The university mathematicians and the mathematics teachers at Desert View High School generally agreed that when they first began to develop projects, they had no real intention to address the *Curriculum Standards,* although the mathematics teachers did have a greater familiarity with the *Curriculum Standards* than did the university professors. One teacher commented as follows:

> *When the* Standards *first became available to us in published form, people had a lot of anxiety and kind of set them aside and said, "We'll see what we can do but we don't expect too much." It seems like it was not until a year ago [1992], after we had a successful program with writing and projects, that we started looking back at the Standards. We realized that our program was aiming our students toward many of the Standards, and it was an affirmation that we must be doing the right thing ... but we never thought of our program as a way of approaching the* Standards.

Like other sites in the R^{3}M study, Desert View illustrated the validating, rather than the generative, influence of NCTM's *Curriculum Standards.* A grant was awarded to the local university by the National Science Foundation (NSF). Part of this grant was directed toward a collaborative partnership with the local schools. One teacher explained the NSF project's facilitative effect on his understanding of the *Curriculum Standards:*

> *A lot of the points in the* Standards*—problem solving, cooperative learning, communication—were at first overwhelming, even if you agreed with them. Now that I have implemented many of them in class through this project, I can see that they are not too difficult.*

As the documenters explain,

> *Perhaps, as another teacher advised, it was not productive to focus on the horse-and-cart question regarding the projects and the* Standards. *What mattered, she asserted, was that the goals and results of both were similar and mutually reinforcing: asking students to build models; to deal with real world phenomena; to communicate mathematically in both written and oral form; and to engage in mathematics as an integrative experience in which concepts are not isolated from each other and students are encouraged to draw upon multiple strategies in collaborative fashion.*

The successful implementation of mathematics projects at Desert View led to some changes in teaching practice, as suggested in the *Professional Teaching Standards,* particularly as teachers began to see their role shift to a more facilitative one—one in which students more frequently engaged in collaborative-group activity—during class time spent working on the projects. Teachers reached out more and more from their typical teaching practice until they felt a degree of comfort in working with the projects, although they expressed a sense of security in knowing that they could return to a more teacher-directed mode on nonproject days. Increased comfort in assuming more facilitative roles, in turn, led the teachers to consider other ways in which the *Professional Teaching Standards* might give new direction to their mathematics teaching.

For Desert View teachers, a desire to validate local mathematics curricula was an important reason to consult the *Curriculum Standards.* Later, as teachers began to take a deeper look at the document, there was a growing consensus that planning for change should follow even more closely what teachers perceived the *Curriculum Standards* to say.

The students attending Scottsville High School traditionally scored among the highest in their state on the PSAT, SAT, and other achievement tests. Through the tireless efforts of the school's mathematics supervisor, technology had become the focus for change in the school's mathematics program. State-of-the-art computer facilities were available for all students to use in conjunction with their mathematics classes. Despite all this, documenters reported that they saw little evidence that the mathematics teachers were committed to the vision of mathematics or pedagogy as portrayed in the *Standards* documents, and the mathematics supervisor did not seem to be motivated directly by the *Curriculum Standards.* From statements made by teachers and other essential personnel involved with the mathematics program, the *Curriculum Standards* appeared to be lending credibility to changes that had already occurred. According to one teacher,

Personally, I don't think—I read it—but I don't go through my lesson plans or a section or chapter of the book and say, "Oh, there was one of those goals" or "Oh, I'll get that copy right now." Its kind of like, our curriculum that we have, if we do the curriculum, it was going to cover those things we are looking for.

The single add-on innovations introduced at Desert View and Scottsville raise interesting questions with respect to teaching practice. What impact do these add-ons have on coherence across the mathematics curriculum? For example, how deep is the implementation of computer use in each mathematics class at Scottsville? How consistent is the application of technology among different teachers in the school? There were indications that the role of the mathematics teachers was becoming more facilitative at Desert View because of the increased use of mathematics projects. Would other add-ons also result in similar shifts in instructional practice? What are the implications of such add-ons for the coherence of the mathematical content?

Teachers at Desert View looked at the inclusion of projects in their mathematics classes as only the start of implementing the *Standards* documents. Their future plans included ever increasing and closer alignments to the documents. The department chair at Scottsville and the principal shared the perception that their mathematics program was in complete alignment with the *Standards* documents. Documenters visiting Scottsville viewed the teaching practice there, in spite of grouping arrangements in some classes and the use of technology, as being somewhat traditional and not necessarily following suggestions in the *Professional Teaching Standards* (NCTM 1991). When teachers said that they were meeting "all the Curriculum Standards," their rationale seemed to be that they were using a textbook series that claimed to be Standards-based. The advanced state of technology in the school seemed to be the basis for another teacher to consider their mathematics program as "exceeding the Standards." And a comment by the principal seemed to reflect the taken-for-granted impact of the *Standards* documents:

Well I don't think there is any question that the NCTM Standards *have been implemented, to what degree I'm not certain, but I would be surprised if the NCTM* Standards *weren't almost completely implemented into our program because I know the mathematics supervisor not only philosophically is in tune with them, I understand that he contributed in many ways into their development.*

Scottsville teachers perceived that their program was an exact match with the Curriculum Standards. Their comments indicated general perceptions of completeness among teachers: "Our courses were built with the *Standards,* that is how they were designed"; "The curriculum that we have here is all within the Standards"; "We exceed all of those Standards";

and "If you look at any one of our courses, or came in and observed our teaching, you could just go down the list of your Standards and be assured that we are right there."

Teachers at Deep Brook saw the Standards implementation as being connected closely with the efforts of a visionary principal who emphasized the need for a different kind of mathematics program even before the *Curriculum Standards* was published in 1989. According to one of the R^3M documenters, this principal

> *was the major catalyst for mathematics reform not only in the Deep Brook School but throughout the district. Through his efforts, professional-development activities were a priority for the past eight years prior to the site visits. He routinely volunteered to model appropriate mathematics instruction at all grade levels within the school, led workshops on mathematics programs, and was the vehicle through which the teachers became knowledgeable about the Standards.*

Through his encouragement and example, teachers, in turn, assumed leadership roles for bringing about the Standards-like curriculum at Deep Brook Elementary School. Several of the teachers regularly conducted mathematics workshops throughout the region to share their insights about how mathematics should be taught.

Teachers in the school readily admitted that the basis for their curriculum was the *Curriculum Standards* and that their local mathematics curriculum's themes were problem solving, connections, communication, and mathematical reasoning. The documenters visiting Deep Brook characterized the learning environment as being "shaped by the types of mathematical tasks presented and the ensuing discourse in which the students became engaged. Teachers constantly tried to perfect their skills by developing and integrating worthwhile mathematical tasks into their instruction."

Teachers at Deep Brook regularly attended workshops, in-service meetings, and mathematics conferences that dealt with ways to interpret the main themes from the *Curriculum Standards* into their classrooms. Teachers were convinced that their mathematical vision was closely aligned with the *Curriculum Standards,* a vision that developed before the publication of the *Curriculum Standards.*

A professional collegiality existed at Deep Brook, where collaboration among teachers was the norm, not the exception. They found that existing textbooks failed to meet their educational goals in mathematics and, over several years, generated a series of problems, projects, and related activities. They believed that a lack of dependency on textbooks forced them to become more collaborative with one another and more creative in their teaching. They recognized the need to engage in frequent planning sessions to develop and maintain what they believed to be an energetic and viable mathematics program.

The four examples cited previously share similar characteristics. One striking difference between Deep Brook and the other three sites is the length of time that local efforts for developing a Standards-like mathematics curriculum had been in place. At the time of the documenters' visit, these four sites had been involved in change efforts from two to eight years. Desert View, the most recent entrant into the arena, was beginning its third year. Deep Brook had had the longer experience and had been involved in the process for eight years. A tendency toward homeostasis that was witnessed at Manzanita, Desert View, and Scottsville did not seem to characterize the teaching practice at Deep Brook. Or did it? How long does it take to implement a Standards-like mathematics curriculum, one in which teachers do not return to a teaching practice that is seen by participants as inappropriate? Will local communities allow teachers sufficient time to naturalize this process of changed instructional practice? What factors, besides time, seem to influence how—and how much of—the *Curriculum Standards* is implemented?

CONCLUDING THOUGHTS

The intent of the R^{3}M project was not to hold up sites as "models" of excellence but rather to provide descriptive illustrations of how the *Curriculum Standards* was being interpreted at various sites. Yet teachers and other curriculum planners wanted a model to follow. There were great differences in the elements of the message of the *Curriculum Standards* that became the central focus within the various sites.

Ball (1992) warns that the mission of the *Standards* documents might stall or be diverted if the Standards are implemented in some mechanical way. The R^{3}M documenters found that the strongest influence on the direction that a Standards-like curriculum would take was the local context in which change in mathematics teaching and learning was occurring. For example, providing classroom sets of calculators or purchasing manipulatives for classroom use may give the impression that changes are being made when, in fact, teachers may have made no changes in the mathematics content or in the way they teach.

Some issues that emerged from the data are worth noting. Within the R^{3}M sites we saw evidence of enormous energy, reflection, and resources being devoted to mathematics education. Classroom teachers, administrators, school board members, and community members displayed a dedicated commitment to improving the quality of mathematics instruction.

What the implementation looked like at the different sites was influenced by the length of time that focused efforts to change had been under way. At the time of our visit, Deep Brook Elementary School had been actively engaged in reform for eight years, the longest time of all the sites in the study. The program at Deep Brook was a good example of the transition from teacher-directed to teacher-facilitated student learning. Most sites, however, had been in the process of mathematics curricular reform for one to three years.

Teachers and others interested in aligning their mathematics programs with the *Standards* documents often ask, "How long does reform take?" Since change does not usually follow a linear model, this question does not have an easy answer. Programs were frequently characterized by their uneven developmental stages. In some instances, documenters witnessed reform efforts that had an early resilience that was followed by setbacks and backsliding. Others were found to start with only a few changes and then pursue a progressive pattern that included more teachers and greater changes. Thus, the complexity and uniqueness of the local context made it difficult to predict how long it would take for implementation of new perspectives on mathematics education to occur or what form it would take. In the example of Deep Brook, teachers believed that their efforts to improve would necessarily need to continue indefinitely. Because local perceptions of the completeness of the implementation differed, it is difficult to predict the tenure of such a process. We did find in several schools that teachers needed about three years to start feeling comfortable with the changes that were taking place, but at each site, teachers still thought that they had a long way to go in their efforts to improve mathematics and science education.

Teachers, schools, and school districts contemplating efforts to improve their mathematics program may wonder what sort of curriculum documents might be needed. We found that at the district and school levels, curricular products ranged from sketchy curriculum guides to a very complete mathematics curriculum document. Robitaille and colleagues (1993) identify three types of mathematics curriculum: the *intended* (or adopted) curriculum, the *implemented* curriculum, and the *actualized* (or realized) curriculum. It is not uncommon for the intended curriculum and the implemented curriculum to be quite discrepant (Ball 1990). Generally, we found that teachers rarely used the intended curriculum of the state or district levels in preparing their implemented curriculum. Assessing student learning was not a part of the R^3M project, so we were unable to determine if disparity existed between the implemented and actualized curricula in the various sites.

Long-range planning was instrumental in changing the mathematics curriculum at the district and school levels at some of the sites; at others, it was not. Where long-range planning occurred, attempts were being made to align mathematics curricula with state frameworks and NCTM's

Curriculum Standards. Curriculum developers in some schools considered the *Curriculum Standards* a mechanism to validate existing mathematics curricula. Deep Brook was the only example of curricular change in mathematics that had been long term and ongoing. The connections were better between intended and implemented curricula at those schools and districts where long-range planning occurred; this was particularly true at East Collins and Green Hills. The other sites were still very much in transition from earlier teaching practices, and the homeostasis of change was much more in evidence and less clearly defined.

Large-scale surveys, such as the National Science Foundation's 1993 Survey of Science and Mathematics Instruction (Weiss 1994) and the Third International Mathematics and Science Study (TIMSS) (U.S. National Research Center 1996), indicate that no prevalent curricular structure for grades K–12 mathematics instruction exists in this country and that Standards-like curricular models are still relatively infrequent at the school and classroom levels. The perception of general acceptance of NCTM's *Curriculum Standards* as a mathematics framework for the United States is predicated on the almost universal use of the *Curriculum Standards* in developing state frameworks in mathematics together with an acknowledgment by teachers of their awareness of current mathematics education reforms (Peak 1996). Although a general awareness of the *Curriculum Standards* may exist, when we break down this awareness to specific pedagogical practices (as was done in the Weiss [1994] study), we see evidence that the *Curriculum Standards* is actually interpreted in various ways. Interestingly, researchers conducting the TIMSS found that at the level of classroom implementation, the mathematics curriculum and teaching practice in Japan align more closely with NCTM's *Curriculum Standards* than those in the United States do (U.S. National Research Center 1996).

Answering the question of what it means to implement the *Curriculum Standards* is not an easy task. It is our hope that data from the R^{3}M study will deepen that discussion.

What Will Standards Implementation Look Like in the Next Ten Years?

There is little doubt that NCTM's *Standards* documents have had a significant impact upon the standards movement in other disciplines. Certainly, the *Standards* documents are at center stage in the current debate on the issue of national standards for education in the United States (Ravitch 1995). What role would the NCTM's Standards play in such a process? Will the revision, tentatively called *Principles and Standards for School Mathematics,* reflect a clearer position of balance with respect to fostering conceptual understanding *and* the acquisition of basic skills? Will teachers still feel a need to supplement Standards-driven materials and texts with teacher-developed and commercial materials that focus on the basics?

As a nation, the United States has a public education system that is controlled at the local and state levels. Can we expect local- and state-level policy to determine the direction that mathematics education is to take in the new millennium? In the State of the Union Address to Congress on 4 February 1997, President Clinton strongly emphasized the need for national standards in education but was careful to state that such standards should not be "federal."

NCTM seems well positioned to develop such standards. Student achievement among eighth graders in the United States shows us trailing in mathematics content mastery by about a year in comparison with other industrialized nations. The phrase *a mile wide and an inch deep* from TIMSS characterizes mathematics curricula in the United States. The TIMSS report recognized that the NCTM *Standards* documents offered direction to deepen mathematics content and teaching practice, but it also indicated that instruction as observed in eighth-grade classes rarely matched the guidelines for teaching mathematics recommended in the *Standards.* Will we, as a nation, be able to achieve some common identity in the way that we implement mathematics content and instruction? Should we? Will the revision of the NCTM *Standards* (NCTM 1998), currently under way, better prepare us to meet these challenges? What will it mean to implement the revision? What will implementation look like?

Chapter 2

WHAT MATHEMATICS IS IMPORTANT?

Karen Graham and Beverly J. Ferrucci

My first feeling was that I really liked the idea of making math more meaningful to the kids, but I wasn't ready.... At the time I first started as a K–3 mathematics specialist, they were still doing standardized tests in second grade. It was very, very stressful for the teacher and children. If your children aren't ready for the test, it's even more stressful. My feeling was that I'm going to incorporate as much as I can and still get the children to where they can do the operations that they need to do a good job on the test. That same year they stopped testing second grade. Then I really had to evaluate what I felt was important. Right now I feel that the operations are just a portion of the mathematics curriculum. I feel there are basic mathematical operations that children must know and must know well. I don't believe in anything that would make a child be intimidated to learn mathematics, like timed tests or contests where they have to get up and do something really fast like multiplication, division, that sort of thing—anything that might give a child a negative feeling about themselves in math. When I do this operations sort of thing, I play a lot of games. We play bingo and the kids love that. They don't realize that they're memorizing, but they love to do it.

—Grades K–3 mathematics specialist
Parker Springs School District

"Potatoes, potatoes, everywhere! What are we going to do with them?" asks a first grader as she enters her classroom and discovers potatoes lying on the students' tables and the teacher's desk. "We are going to become potato farmers for a week," replies the teacher. For the next few days the students would be estimating the circumference of potatoes, discussing various methods to weigh them, conducting a survey of their favorite ways to prepare potatoes, graphing the data from their survey, and using money to purchase materials to design potato creatures.

—R^3M documenter
Deep Brook Elementary School

It has been almost ten years since the National Council of Teachers of Mathematics released its first *Standards* document, *Curriculum and Evaluation Standards for School Mathematics* (1989). Since that time, *Professional Standards for Teaching Mathematics* (1991) and *Assessment Standards for School Mathematics* (1995) have been published to expand on the content and vision of the original document, and a revision is under way. States have developed standards and frameworks; textbooks advertise consistency with the *Standards* documents, and teacher education efforts are using the *Standards* documents as a core (Cohen 1995; National Council for Accreditation of Teacher Education 1994). Nonetheless, we are only beginning to accumulate a knowledge base about the process and extent of reform in mathematics education as it is occurring in this Standards context.

The *Standards* documents and Standards-based state frameworks describe new goals for the teaching and learning of mathematics. These goals include helping students make sense of traditional topics; offering opportunities for students to explore more novel topics, such as probability and statistics and areas of discrete mathematics; advocating the design of more technology-based courses; and helping students to reason with, communicate about, and apply mathematics in real-world contexts (NCTM 1989; Rosenstein, Caldwell, and Crown 1996; New Hampshire Department of Education 1995). The *Curriculum and Evaluation Standards for School Mathematics* (1989) recommends mathematical content in the three grade-level groups K–4, 5–8, and 9–12. It discusses which topics should receive increased emphasis and which should receive decreased emphasis. The document presents guides for teachers or schools to follow as they develop their own curricular emphases. In addition, Standard 1 of the *Professional Standards for Teaching Mathematics* (1991) describes "worthwhile mathematical tasks" consistent with these goals. According to this document, such tasks should

> Engage students' intellect; develop students' mathematical understandings and skills; stimulate students to make connections and develop a coherent framework for mathematical ideas; call for problem formulation, problem solving, and mathematical reasoning; promote communication about mathematics; represent mathematics as an ongoing human activity; display sensitivity to, and draw on, students' diverse background experiences and dispositions; [and] promote the development of all students' dispositions to do mathematics. (P. 25)

Although the documents appear to articulate a clear vision of mathematics for the grades K–12 setting, questions remain about how the mathematics content enacted in schools and classrooms compares with this vision and how teachers perceive the content recommendations. Within the R^{3}M sites was a range of mathematical choices and emphases, and pedagogical issues were generally addressed more directly than mathematical ones. This chapter focuses on how some of the R^{3}M sites dealt with the mathematical issues at the heart of curriculum change. We use examples from the R^{3}M sites to illustrate what kinds of tensions emerged around the "basic skills" balance; what the schools and teachers focused on, and took from, the *Standards* documents to define the "mathematics" component of their reforms; how teachers were interpreting problem solving and communication; what role the teachers' content knowledge had in the change process; and how sites dealt with the issues of making mathematics relevant and meaningful for all students.

WORKING TO DEFINE THE "MATHEMATICS" COMPONENT OF REFORM

The role of mathematics content in reform initiatives is a crucial component at all levels. By the "mathematics component" of reform we mean the mathematical content and processes that teachers, specialists, and school administrators believe are important for students to learn. It includes how that mathematics content is organized and what mathematical processes are most emphasized. It is important to note that to date, the community has done a relatively thorough job of articulating the pedagogical changes that are giving definition to mathematics education reform, but the mathematical shifts that are being undertaken have received less emphasis. So, although the recommendations have been made in both areas, in *what* mathematics is taught as well as *how* it is taught, it appears that the how has received the most attention and may have been easier to communicate initially. The R^{3}M sites reflect this struggle. Although several of the sites expressed the desire for the mathematics curriculum to move beyond the notion of mathematics as computation, they had trouble articulating what the specific changes would be or how they would be described mathematically. Examining the "mathematical visions" of several sites presents a context for discussing issues about the mathematical dimensions of grades K–12 reform worthy of further exploration.

Developing a Philosophy for Mathematics Teaching and Learning

The Perfect Situation for Mathematics Learning (PSML) project came to life as a partnership among industry, East Collins High School, and the state department of education. Planning for the PSML project began in January 1991 and represented the combined efforts of teachers, administrators, and mathematics professionals, who worked with business leaders to create a different mathematics learning environment. PSML was created with assistance from the state's Coalition for Excellence in Mathematics Education, and the team envisioned the project as being firmly grounded in the *Curriculum and Evaluation Standards for School Mathematics* (NCTM 1989).

The mathematical and pedagogical visions of teachers at East Collins is set forth in a brochure describing PSML:

> In a PSML classroom, students learn to solve problems through a hands-on approach, using both modern technology and everyday materials. Abstract principles are made concrete as students conduct lab experiments—generating data, posing problems, and employing mathematics principles to arrive at solutions. Computers, graphing calculators, journals, and teamwork strategies are used to integrate mathematics concepts with technology and verbal skills. Reading and communication skills are stressed in completing individual and group tasks, much like assignments in the workplace. Teachers work with students on *how* to solve problems—rather than dictating solutions and assigning repetitive sets of calculations.

The need that drove PSML was the recognition by mathematics teachers that a large number of East Collins students were choosing to enroll in nonalgebra mathematics courses in their ninth-grade year and that these students were not as challenged mathematically as they might be. The chair of the mathematics department indicated a readiness for mathematics reform at East Collins High School (ECHS): "We just felt as a math department that there was a better way and so that feeling made us very receptive. So when we had this opportunity, we just jumped at it. It was great timing." Implementation of the project began at ECHS and East Collins Middle School in fall 1991 with prealgebra and algebra 1 classes. PSML expanded to include algebra 2 during the 1992–93 school year and Euclidean geometry in the 1993–94 school year. Funding from industry helped in the purchase of some textbooks, supplementary materials, manipulatives, a computer for each mathematics classroom, and summer staff development meetings during the first two years of the project.

How did this curriculum project differ from those in which East Collins High School had engaged in the past? In particular, what differences existed for the students in the district? The R^3M documenters commented that during their visit they noticed calculators in the classrooms at all grade levels. When they questioned this presence, they were told that the district had purchased a classroom set of twenty-five calculators and an overhead projector for everyone who taught mathematics in the district. The high visibility of these calculators served notice to teachers, parents, and students that the district did indeed expect changes in the teaching and learning of mathematics at all grade levels. The following is a summary of the classroom observations at the high school level.

At East Collins High School, the R^3M documenters observed a geometry class working cooperatively on a project involving the application of geometric constructions. The task was to design a wrapper for a six-pack soda carton with a one-inch overlap on the basis of the measurements of one can of soda. Various supplies were assembled for the students to use: soda cans, customary and metric rulers, scissors, calipers, TI-30 scientific calculators, wrapping paper, glue, tape, markers, and paper for planning purposes.

Before starting on the task, the students were presented with a rubric that would be used to score the project. Once they were satisfied that they understood the scoring system, many students started by measuring a can. Other students had questions: "Is the overlap on the top, bottom, or side of the cans? Is the wrapper a single piece, or can it have seams? Should the wrapper fit snugly around the cans?" After these questions were addressed by the teacher, the students divided into pairs and began the task. Some of the pairs worked in metric units and others, in customary units. Calculators were used for conversion between scales. Students who finished the project early were encouraged to decorate the wrapper or develop commercials. All the students appeared to be engaged in the project, and no one seemed to be off task. When all students were finished with their designs, they were asked to self-assess their work by using the rubric that the class had previously discussed. The final assessment challenge was for them to take their wrapper and try it out on the six-pack of soda sitting on the teacher's desk. As they discussed this lesson, we asked the teacher if this discourse was typical of her classes. She responded no, since this lesson was planned for the ninety-minute period that is part of the schedule only once every four weeks. Components of the lesson—group work, hands-on materials, self-assessment, and calculators—were, however, part of her daily lessons, which was evident in the students' behavior in this type of investigative setting.

In an algebra class, students were using TI-81 calculators to investigate patterns in graphing rational equations. The teacher described the intent of the lesson as using patterns to see the relationships of transformations on a given function. She cited the pattern-recognition strand of the new mathematics curriculum as part of the rationale for the lesson. The teacher started the lesson with a quick review of the steps needed to graph a function. None of the students appeared to have any difficulty with this technical part of the lesson. The students then broke into groups of three or four to complete the assigned activity and to discuss the patterns they discovered when comparing the graphs of the given functions. The teacher asked the groups to respond to such questions as "What was your prediction for the graph for a given function?" and "How did you adjust your prediction when it was inaccurate?" Students were also asked to write conclusions about their observations. The teacher noted that the exercises in this lesson had been designed by the department chairperson and shared with other faculty members. She thought that this was important in encouraging other teachers to incorporate the graphing calculator into their lessons.

During a group meeting with the documenters, the high school faculty spoke about creating problems relevant to the students. One teacher described a lesson that failed. She noted that the students were given information relevant to establishing a radio program of music, news, and advertisements. She was the station manager and wanted a report. The students in groups of two or three were to gather the information they needed, organize the data, and decide on how best to present the information. The teacher explained that one difficulty was that she had not factored in the amount of time needed for the students' preoccupation with selecting the type of music and, indeed, with choosing exactly which songs. This situation raises interesting issues about factors that teachers need to consider when designing open-ended activities. Had the selection been a homework assignment—to bring in that information—the lesson might have been better. The choice of topic does illustrate the faculty's concern with tying mathematical experiences to the real world of the student. It also demonstrates a departure from arithmetic as content, which would have been in place before the commitment to the new curriculum.

Before reading further, you might find it helpful to reflect on the following questions in the context of the foregoing description of PSML. How do the classroom observations compare with the PSML vision? What are the most salient features? How would you describe the mathematical vision of your district? Would you describe the actual changes that have occurred as mathematical or pedagogical? The next section explores a vision of mathematics as problem solving and communication.

Mathematics as Problem Solving, Communication, and Reasoning

The preceding section discusses one site's development of a philosophy for mathematics teaching and learning and how that philosophy is translated into classroom practice. A review of the R^{3}M documenters' summaries of each R^{3}M site's mathematical vision statements reveals a strong presence of the first four Curriculum Standards: mathematics as communication, mathematics as problem solving, mathematics as reasoning, and mathematical connections (NCTM 1989). A teacher at one of the R^{3}M sites explained, "We're trying to make a more global approach and to bring all the different things together, rather than math skills as such. Teaching kids to think and reason." Representative of the teacher's vision at this site were comments about the importance of communication, the need for mathematical reasoning, and the importance of problem solving. One teacher explained, "I'm working with the strands of the [*Curriculum*] *Standards* and am currently emphasizing patterns through the use of pattern menus, summary statements, and self-assessment." The principal at this same site commented, "We need to teach children to think."

Documenters visiting one of the R^{3}M high school sites summarized the mathematical vision and subsequent curriculum changes as "mathematical reasoning and developing communication skills." When asked to describe what kinds of learning the students experience, the faculty at this site responded with "hands-on, group problem solving, critical thinking, confidence to get started on any problem, willingness to work on mathematical problems, risk taking, making connections between ideas in mathematics and mathematics and the real world, making use of multiple representations, use of appropriate tools to support problem solving, calculators, computers, other manipulatives."

Another high school site built reform efforts almost exclusively around the development of a "projects" approach to instruction. The documenters stated that through "projects, themes, and emphasis on writing, the mathematics teachers appear to hold a vision of mathematics as an integrative experience in which concepts are not isolated from each other and students are encouraged to draw upon multiple strategies and disciplines when solving real-life or relevant mathematics problems." Another interesting component of this particular site's mathematical vision is that teachers want to construct opportunities to communicate mathematics with one another. They see this collaborative effort as a means of strengthening their own mathematical understandings. Part of the vision held by teachers is directed toward motivating students to find mathematics interesting.

The following questions might be helpful as you reflect on the documenters' stories in the next two examples of schools at other levels. Is "math is fun" one of the important messages of the *Curriculum Standards*? Should motivating students at any price be a mathematics teacher's primary goal? Why are communication and problem solving important goals? What about the classrooms in these sites fostered communication and problem solving? What mathematics was being studied in addition to the processes? On which—the content or the process—do you think the students were more focused? On which do you think the teacher was more focused? What are the strengths and weaknesses of the approaches taken by the teachers in these classrooms? How do the two sites compare in their interpretation of communication and problem solving? How does a teacher structure a lesson that emphasizes communication and problem solving but does not give the impression that any response is acceptable? How do we achieve a healthy balance between the process of doing mathematics and actual mathematics content?

Bedford Middle School

The documenters reported the following:

> *As we walked in to observe a seventh-grade class, the teacher turned to greet us. "Hey, come on in. We've been expecting you." We threaded our way to the back of the room through a maze of armchairs arranged in tight rows and settled down to observe. It was a hot, sunny day, but the room had no air conditioning and only a couple of small windows that could be opened for fresh air. Twenty-four Hispanic students, working in pairs, turned their heads briefly to inspect us, then returned their attention to their task.*

"Miss? Do we need to write our observations in our journals?" "José, would you please tell Ana and Lucia what the directions for this assignment were? They were still worried about finishing the last task when I explained this new one to the class."

Tacked up around the classroom were humorous posters about proper behavior, a poster illustrating a Math Explorer calculator, and another one about tessellations. The only bulletin board in the classroom, labeled "Math Logic," boasted such mathematics-related sayings as "The only way to learn math is by doing it" and "There is both power and beauty in math." A hand-printed list of "Goals for All Students," which was essentially the goals from the NCTM's Curriculum Standards, *also hung from the board. The equipment in the room included a television and a video recorder, a Macintosh computer with a printer on the teacher's desk, and an overhead projector. Numerous buckets of manipulatives were on the windowsill beside the teacher's desk, and the bookshelves and cupboards were overflowing with papers, books, and materials.*

Over three days, the documenters observed the classes of all five Mathematics Teaching Project (MTP) teachers at Bedford Middle School (BMS). The MTP was a districtwide effort aimed at preparing all students to experience algebra in grade 8 or 9. The goal was to introduce preliminary algebra-based activities throughout the first eight grades of school. The five MTP classrooms looked about the same, except for one where a lot more student work was displayed on the walls. Lots of teaching materials were evident in all the rooms. Many of the lessons that they observed used manipulatives. They saw activities with algebra tiles, attribute blocks, and paper folding. Teachers seemed to bring these materials into their teaching on a daily basis, and students used manipulatives with familiarity. Two ideas seemed most prominent in the teachers' collective vision of what is important in mathematics teaching and learning: motivation and communication.

Motivation

A message heard over and over from the BMS teachers, as well as from the school's administrators, was that mathematics must be fun. If students are not having a good time, they are not engaged and learning cannot take place. Therefore, the primary goal of the BMS mathematics teachers was to motivate their students and engage them in the lessons they were teaching. A teacher at BMS gave the following response to the questions "How would you describe a successful teacher?" and "What makes your class special?"

[A successful teacher would be one] who could motivate, who knows how to reach all the different students, address all the different learning styles—visual, verbal, auditory, whatever. What

makes my class special? The kids. Their intelligence level. Their overcoming their obstacles. The fact that they believe they can do it. The fact that it's fun to be in there. The fact that they get to do lots of different things. It's not all just take notes, do twenty problems, and grade it. I mean, it's not boring. The fact that they're allowed to share. The fact that they're not isolated. The fact that they're allowed to be loud and noisy if they want to. The fact that they can go off on Beavis and Butthead for a few minutes. I will pull them back into the class. The fact that math is great, and it's fun, and they're being able to learn that.

In all the classes observed, the students seemed to be engaged in the lessons that the teachers were presenting. They listened when their teacher talked; they responded to questions; and they worked diligently at their seats, focusing their attention on the tasks assigned. All the teachers used manipulatives or overhead transparencies to illustrate the ideas they were presenting. They led whole-group discussions or orchestrated class presentations; they posed questions, challenged students, and suggested conjectures to be investigated. The students for their part asked questions to understand the lesson or task presented, followed their teachers' suggestions for exploration, and shared their thoughts with their peers.

Communication

The BMS teachers said that they valued students' being able to talk about mathematics, which was borne out through their efforts to have students explain what they were doing or what they understood about the assignment. The main approach used by the teachers to facilitate communication was to have students work in small groups. Most often, students worked on all assignments with the classmate sitting right next to them. Student partners also served as monitors for each other, which freed the teacher to circulate around the room and facilitate the work of the students.

The kids, when they get here in the class, they know they're going to be challenged, and they know that they've got to think. There's just no way around it. And so among the things that I do in my classroom, I like to incorporate activities where they communicate. I like them to be able to explain, to justify [what they are doing]. In my classroom I focus a lot on problem solving, on having my students communicate and reason. So for me the Curriculum Standards *has really placed an emphasis on that. I know that communication is a major one for me. As I alluded to earlier, I have my students do a lot of communicating, whether it's verbal or it's on paper.*

—Mathematics teacher
Bedford Middle School

Deep Brook Elementary School

At Deep Brook, one of the elementary school sites, the documenters observed that the "principal and staff are attempting to form a strong mathematical community by promoting problem-solving skills through communication, making connections, exploring patterns, graphing data, explaining strategies, and encouraging the use of multiple representations." Documenters at this site reported that a problem-solving environment was accomplished by having the students interact with mathematics in the following ways:

- Students developed enough confidence in their understanding of mathematics to feel comfortable estimating, hypothesizing, exploring, and verifying.
- Correspondingly, teachers fostered an environment in which students are risk takers and feel secure in volunteering responses without fear of ridicule or censure for incorrect answers.
- Students recognized the importance of mathematics in their everyday lives.
- Students communicated their ideas in a mathematical format.
- Students constructed and explored their own mathematical connections.
- Students were encouraged to articulate and defend their problem-solving strategies.

Effective mathematics teaching, according to the Deep Brook School's administration and staff, occurs when teachers are allowed to take risks and to experiment with new techniques and materials. The Deep Brook teachers have helped their students develop enough confidence in their understanding of mathematics to feel comfortable estimating, hypothesizing, exploring, and verifying.

The students at Deep Brook recognized the importance of mathematics and willingly spent a great deal of time working on mathematical tasks. They knew how to select appropriate heuristics, collect data, make appropriate adjustments, verify the results, and continue this cycle if needed. As one fourth grader noted,

> *My favorite strategy is looking for a pattern. It is easy and makes you feel good when you see it. It makes math fun! I also like the strategy of making a list of all possibilities. When I was younger I used to like guess and check. I don't like that anymore because it sometimes takes too much time, but it's okay when the problem is very easy.*

Deep Brook students were encouraged to articulate their mathematical thinking both orally and in writing. They used mathematics journals to document their emerging understanding of mathematics concepts and strategies. Even kindergarten students maintained a class mathematics journal in which the teacher transcribed the students' oral explanations of strategies. Using the journal coincided with the broader emphasis on fostering in students the ability to verbalize mathematics strategies and engage in dialogue within a context of problem solving. Excerpts from these journals follow:

> *Today's Problem* (grade 4): 2 apples weigh the same as a banana and a cherry. 9 cherries weigh the same as a banana. How many cherries weigh the same as 1 apple?
>
> *What strategies did you use to solve it?* First I drew 2 apples = 1 banana and 1 cherry. Next it said 9 cherries = 1 banana. The fruit that didn't belong with the apple and the cherry was the banana. So I exchanged 9 cherries with the banana. Next I put five cherries in each column. Finally I figured out that 5 cherries equals one apple.
>
> *What patterns or relationships did you find?* Exchange the cherries with the banana.

> *Today's Problem* (grade 4): Pretend that there are 65 students in our school with a ratio of 3 boys to 2 girls (3:2). How many boys and how many girls are there?
>
> *What strategies did you use?* I drew 3 boys to every 2 girls and I counted up to 65 'cause there were 65 students. The answer I got was 39 boys and 26 girls.
>
> *What patterns or relationships did you find?* If you count by fives it will go into 65 evenly.

During an activity in a kindergarten class, students were asked to give a reasonable estimate of the number of seeds in a box. (Note: The box contained forty-seven seeds.) One student described his thinking process: "Well, I know that there is more than one seed in the box. And no way is there a million seeds in the box! One hundred sounds like a good guess. Yeah, I like a hundred." He paused for a moment to listen to his classmates' guesses before he decided to change his estimate: "Okay, I want to change my guess. I can fit about ten seeds in my hand so I would guess that I can fit four hands in there. So my guess will be forty. No, make that forty-one."

TENSIONS SURROUNDING THE INCLUSION OF BASIC SKILLS

What is the ideal balance between the development of skills and the development of concepts in mathematics teaching? Although ten years have passed since the release of the first *Standards* document, this question continues to be important and sparks heated debate in many circles. The list in the *Curriculum Standards* that outlines the content to receive decreased emphasis is often misinterpreted as a call for the elimination of skills from the curriculum. Current backlash movements in several communities call for more emphasis on skills and computation and less emphasis on problem solving and reasoning. Evidence from classroom observations and curriculum guides suggests that at the time of the R^{3}M visits, the predominant emphasis was still on computational skills at many of the sites, especially at the elementary level. The following observation is typical of what documenters reported:

> *Many of the teachers were focusing on computationally oriented activities, such as place value, the monetary value of a combination of coins, equivalent fractions, and long division, although there were a few teachers who were trying to teach problem-solving skills to their students. The kindergarten teachers were trying to teach their students to look for patterns. A fourth-grade teacher was having her students make predictions and use multiple approaches to arrive at the answer. A fifth-grade teacher asked her students to find the largest power of 2 which they could display on their calculators.*
>
> —R^{3}M documenter

Several of the elementary sites are clear about their view that mathematics is more than "just numbers," although they struggle with the need to cover the computational aspects of mathematics at this level. One documenter reported the following:

> *There are strong feelings on both sides of this issue. A few of the teachers are trying to change the way that they teach mathematics due to their participation in workshops and in-service activities. Other teachers would like to learn how to make changes, but the majority have no desire to change their mathematics program. They are satisfied stressing computational facility and want to remain in a drill-and-practice atmosphere.*
>
> —R^{3}M documenter

The opening vignette in this chapter presents a teacher struggling with these issues and suggests that one reason for the struggle is the presence of standardized tests. At Southland Elementary School, teachers used personal interests as their vehicle for connecting students with mathematics. Along with this commitment at Southland to give students real-life experiences with mathematics, the documenters noted that as they observed classes and spoke with teachers, an apparent emphasis on building computational skills through practice and memorization surfaced. Teachers noted that they had "studied" the *Curriculum Standards* and valued the decreased attention to rote skill development. However, they still had to administer a basic-skills achievement test on a schoolwide basis. The teachers agreed that assessment was a large concern and that it was just beginning to be addressed at the school. Until that issue was solved, the teachers believed that they could not ignore "basic-skills practice." Teachers at Southland and elsewhere often appeared unwilling to risk changing their practice away from basic skills and computation for fear that their students might perform poorly on standardized tests. Teachers often felt pressure from colleagues in other grades who demanded a certain degree of proficiency with symbolic manipulation of both the arithmetic and algebraic varieties.

Johnson (1995) in his cross-case analysis of the secondary school R^3M sites, reported that the teachers at Pinewood High School found it difficult to de-emphasize the time spent on learning basic skills because of low test scores in the school. Scottsville, a school where students traditionally scored high on standardized tests, also had a mathematics department committed to ensuring that students mastered important mathematical skills. Teachers at both sites thought that time spent on conceptual development borrowed valuable time from skills development and prevented teachers from covering needed material. In constrast, teachers in the three other high school sites found that they were able to make the transition to a focus on mathematical concepts and processes. At Green Hills, alternative statewide assessment in mathematics was encouraging or driving instructional practice that encouraged conceptual development. Teachers at East Collins and Desert View indicated that they felt little pressure from standardized testing, and they readily adapted to a process-oriented curriculum. Teachers in those two schools believed that skills acquisition followed naturally if mathematical concepts were emphasized.

Other sites, such as that of the teacher in the opening vignette, attempted to embed the coverage of skills in a context. A visit with a fifth-year mathematics specialist in the Parker Springs School District produced the following story.

A Mathematical Moment

As we entered the room, we noticed our names written on the chalkboard, welcoming us and informing the students that we would be visiting the class that day. The third-grade teacher was working with the class on a variety of multiplication facts, which on first inspection seemed to be low level and unrelated. As she continued with the problems, it soon became apparent that the students were giving her multiplication facts based on their names: the number of letters in a student's first name times the number of letters in a student's last name. As students gave their mathematics problems and the teacher wrote them on the chalkboard, other students checked them by using calculators, pencils and paper, or their heads. When all the students had given a problem, we were introduced to the class. One child said aloud, "I wonder what's the value of their names," and pointed to our names on the chalkboard. The teacher replied, "What a good question! Let's figure it out."

As it turned out, the problem was not simple because the class had two ways to calculate the value of a person's name. The first was to multiply the number of letters in the first name by the number of letters in the last name, as mentioned above. The second way was to give each letter a value (a = 1, b = 2, c = 3, ..., z = 26) and then add them. The class at times struggled with which way to figure the name values, but the teacher let them work through the situation.

The discussion then returned to the children's multiplication sentences. The teacher posed the following question on the chalkboard: "Whose product name has the greatest value? If you think your product name has the greatest value, stand up." Five children stood up. Together the teacher and the students verified that indeed a student had a name product of 99. She then asked if anybody's name product was greater than 99. One student thought that one of our name products might be greater than 99, but he hesitated a minute and retracted his suggestion. The teacher questioned him about why he had changed his mind. The student said that he was thinking that he needed to find only the sum. The teacher then said, "But if you wanted to find the product of all the letters in her name, what would it be?" The students responded that they would need calculators to figure it out.

At that point, students raced for calculators and began to multiply the values of each letter. As students came up with solutions, they checked their solutions against those of their classmates and the teacher. When the teacher and several students reached the same solution, they went around the room, checking the solutions of all the students. At one point, a student compared her answer with the teacher's answer. They were different, so they both rechecked them. The teacher had the incorrect answer; the student had the correct answer. The teacher used this opportunity to discuss the advantages and disadvantages of using a calculator. When all students appeared to have the same solution, the teacher asked them to give all the digits in the number.

Students: *17 962 560*

Teacher: *Agree or disagree?*

Students: *(All together) Agree!*

Teacher: *Now the problem is, What is that number?*

The students tried to read the number. When a student was unsure, the teacher and other students tried to help out. The teacher continued to encourage the students to try and thanked each one for his or her contribution. When one student finally read the number correctly, the students yelled out jubilantly, "Agree!"

—R^{3}M documenter

The teacher in this classroom was not afraid to follow the path set by one of the children. At the time of the first visit, the class had not begun to discuss multiplication. They had just started to look at multiplication facts by focusing on their product names. The teacher saw herself as a risk taker and believed that students' capacity for learning is far greater than most teachers realize. To her, it was not enough just to glance over the question or the solution; she wanted to sustain the discussion and work the problem through until the students were satisfied. The class stayed with the problem, and it appeared that no one was left out of the discussion.

Embedding the learning of skills in a context raises another question about whether the emphasis is still on algorithms, but algorithms embedded in a context. If this is so, have we really progressed very far? The mathematics consultant at one of the middle school sites explained in these words:

We looked at the discrepancy between math in schools being hundreds of years old and mathematics being a living, growing body of knowledge that's been invented fairly recently. We looked in terms of a need for change. I looked at number and I looked at a couple of things that are still real important in number. One is a sense of number and one is being able to use numbers to make sense of situations, so I took them through a thing where I wanted them to [see that because of] the way we learn math, we have no sense of number. I looked at paper-and-pencil algorithms no longer making sense in the 1990s. That what we want is to give kids situations and look at what algorithms they invent based on relationships and how they invent algorithms based on relationships versus our paper and pencil.

—Mathematics consultant
Manzanita Middle School

Does this emphasis move us beyond thinking of mathematics as solely arithmetic? An administrator at the site expressed the dilemma as follows:

You still see some of the computational aspects; that's been a real hard thing to integrate as far as everything always having a purpose, rising out of a purposeful situation. You'll still see artificial situations created to learn the different computational skills that people believe.

—Administrator
Manzanita Middle School

The struggle to know how much emphasis to place on basic skills raises several interesting questions: How would you describe your school's mathematics curriculum? Where would you place your curriculum on a continuum with basic skills at one end and concepts and processes at the other? What is an appropriate balance for your school? Many states are adopting assessments that are compatible with recently developed Standards-based state curriculum frameworks. Is such an assessment in place in your state? Has it had an impact on your curriculum? Would you describe your school's reaction to standardized tests as proactive or reactive? How can basic skills be incorporated into a Standards-based curriculum?

A documenter who visited one of the elementary sites attributed the basic-skills struggle to whether teachers at the site had attended in-service workshops. The documenter wrote the following:

The teachers we observed who had participated in mathematics workshops and/or in-service activities were starting to change their questioning techniques and the types of problems which they were presenting to the class. They encouraged multiple strategies and often concluded the class with a summary discussion of what mathematics had been learned in the class that day. The rest of the teachers were using a computational approach which stressed the mechanics of the problem with little time for reflection or understanding of the basic concept.

—R^{3}M documenter

This observation raises interesting questions about the role of in-service activities in mathematics education reform and in particular about how a teacher's knowledge of mathematics seems to be reflected in his or her practice or attempts at change. The following section explores how one R^{3}M site dealt with this issue.

TEACHERS' CONTENT KNOWLEDGE

The West Redwood School District, one of the largest districts in the country, has an urban focus. The majority of the students are African American and Caucasian, with a few Asian students. The district has been under a desegregation order for thirty-five years, which has resulted in cross-town busing. The district has been revamping its mathematics program since 1983. Ten model schools feature a departmentalized grades K–5 elementary program. Teachers elect to teach either mathematics and science or language arts, and the children rotate through classrooms throughout the day. Two mathematics specialists are involved at the school for grades K–3. These specialists give demonstration lessons in classrooms and plan additional staff development. Besides visits from the mathematics specialists, one day a month the classroom mathematics teachers receive staff development that includes professional reading, mathematics content instruction for teachers, and specific student activities for use in the classroom. These in-service activities are presented by the specialists, the mathematics curriculum supervisor, and the teachers themselves. However, the majority of the mathematics content intended for the teachers is presented by Dr. Curly, a mathematics professor from the local university. The mathematics specialists explained the process:

> *We do have Dr. Curly come in as a consultant in the afternoons, and he will actually do some content-oriented activities with the teachers. Because we don't want them to come in and look for just quickie activities that they can take back to the classroom. We want them to have some mathematics content as well.*
>
> *The teachers want activities ... and most teachers would like to have something to take right back; they don't realize that content is going to help them down the road.... Content is important and the rationale about why they are doing what they are doing is just as important.*
>
> *They keep a journal and then I write back to them and it is very private and we've been doing this. It's our third year doing that, and I think you see an awful lot. How frightened they were at the very beginning—especially with the mathematics content—very frightened and insecure and didn't like it at all, and now they are moving on.*
>
> *It was really funny—we would ask for input from the teachers and when they got started they would say, "I do not like the math, I do not like Dr. Curly, I cannot understand him." So I would tell them [to] give suggestions of what they liked and what they wanted to see. And we ... would have some activity, something for them to take back every time. And so I told them, 'I'm going to take your input ... but remember the main focus of this project is for mathematics—and you will get more comfortable.' And they really have.*

Dr. Curly explained his use of rectangle math in the teacher in-service courses:

> *I can ask whether 3 by 5 is more or less than 2 by 6 without ever saying 15 and 12.*

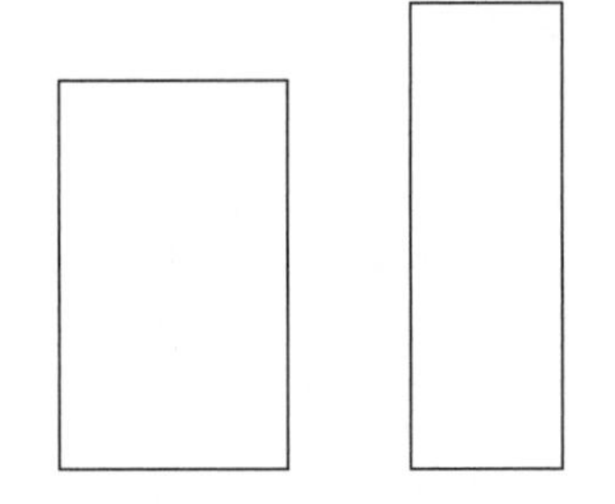

> *And so we played all kinds of rectangle games. For one thing, so you take blocks and you can use whatever manipulatives you have. Let's say you give people seventeen blocks and say, "Make as many different rectangles as you can."*
>
> *The first question is, Does the 1 by 17 count as a rectangle? And some people are very disturbed by that. They don't want to call that a rectangle. And then some people wanted to call it 1 by 17 instead of 17 by 1 or call them different. And so we just opened it up; are they the same or are they different? We voted at the in-service, and most people wanted to call them different because you can make two rectangles with seventeen blocks.... And then we go on.... And we talked about things. Why? For example, why is 3 by 5 equal to 5 by 3? ... I've done this for many years. I've never had a teacher give me a reason other than because they're both equal to 15, and I think that's abominable. They're equal because you can look at the rectangle from this side or from this side.... And I think we really missed the boat by not having very young kids do lots and lots of activities with rectangles and having multiplication introduction.*

The documenters then observed a fifth-grade teacher using rectangle mathematics in her lesson. The teacher asked the students to give "fact families for multiplication and division" for the number 28. The following dialogue took place during the observation, which is described by a R^{3}M documenter.

> *Two students wrote on the chalkboard and read:*
>
> $$4 \times 7 = 28,\ 7 \times 4 = 28$$
>
> $$28 \div 7 = 4,\ 28 \div 4 = 7$$
>
> $$\begin{array}{r}4\\7\overline{)28}\end{array}\qquad\begin{array}{r}7\\4\overline{)28}\end{array}$$
>
> Teacher: *I see that the light for many of you has been turned on, and you can see your fact families for multiplication and division. For each fact family, I want you to turn*

in one piece of paper with both names so I can see what you are doing. [On the overhead projector, the teacher places a 10 by 10 grid. On their desks, students have large-grid graph paper and scissors.] *The week before vacation, we worked on multiplication arrays. We are going to use graph paper to set up multiplication and division fact families. Using this paper, how can I show 8 divided by 4?*

Sharon is asked to draw the problem on the overhead. She walks to the overhead, but doesn't draw anything.

Teacher: *I need two groups of 4.*

Sharon nods but doesn't draw anything.

Teacher: *Can anyone help her? 8 divided by 4 = 2.* [The teacher draws a 4 by 2 rectangle.] *How can we show her that?*

Sharon starts to draw a line, but it's not by the teacher's rectangle.

Teacher: *Not quite. Let me help you. Because of time, I'm going to help you.*

The teacher draws a line dividing her rectangle into two groups of four and explains that she now has a representation of 8 divided by 4 that equals 2.

Teacher: *Let's see if Mary can do one because she's having trouble. 10 divided by 5 equals?*

Mary: *I don't understand.*

The teacher walks over to Mary and talks quietly. She then draws a 5 by 2 rectangle on the graph paper.

Teacher: *How can you do two groups of five?*

Mary starts to draw a different rectangle. The teacher asks for someone to help Mary. Steve says that he understands and walks up to the overhead projector. He begins to draw a new rectangle.

Teacher: *No, on this one* [pointing to the one she drew].

Steve draws a line on the rectangle drawn by the teacher, dividing ten blocks into two groups of five.

Teacher: *Good. We need to do more work on this.*

How does the teacher's knowledge of mathematics in this scenario seem to be reflected in her practice? How does the teacher's lesson compare with the focus of Dr. Curly's activity? The *Curriculum Standards* calls for significant changes in what mathematics should be taught in schools. Many current teachers have not experienced these topics as learners.

How do we help teachers improve their own mathematical content knowledge? What are the most appropriate in-service activities? How can we move away from the "quick activity" focus of many in-service courses and begin to build more sustainable efforts? Given the current backlash discussions, it would seem particularly important to give teachers opportunities to strengthen and broaden their content experiences. A by-product of such experiences would be a mathematical vision that is rich and powerful. Ball (1996, p. 50) states that "the mathematics reforms challenge culturally embedded views of mathematics, of who can—or who needs to—learn math, and of what is entailed in teaching and learning it, we will find that realizing the reform visions will require profound and extensive societal and individual learning—and unlearning—not just by teachers, but also players across the systems."

MAKING MATHEMATICS RELEVANT

We begin this section with a documenter's story:

> *Sitting in the makeshift library of the school, two tables and four bookshelves tucked away in the back corner of a hallway, I began sorting my papers and preparing my questions for my first student interview of the day. As the 5′10″ male eighth grader walked toward me, I smiled and invited him to sit at the table with me. I began the interview with the first question I used to begin all my interviews: "So, what do you want to be when you grow up?"*
>
> *Students' responses to this question usually lead very nicely into my next question: "And how will mathematics help you with your future plans?" But today's interview was not the norm. In fact, it was not like any interview that I had encountered in all the R^3M interviews.*
>
> *"So, what do you want to be when you grow up?" I asked. The eighth grader looked at me incredulously and repeated the question, "What do I want to be when I grow up?" He repeated in a slightly firmer voice, "What do I want to be when I grow up? Man, that's a stupid question!" He continued to stare at me with a look of bewilderment and repeated, "What do I want to be when I grow up? You know Manuel and Lenny are gone now. Both gone in the streets in one year. What do I want to be when I grow up? Man, I just want to still be alive!"*

This scenario occurred at Garnett School, a grades K–8 urban school in one of the poorest neighborhoods of a large metropolitan city. Most of the students lived in the subsidized housing project located across the street from the school. For many of them, the school was a safe haven from the drugs, street crime, and vandalism that filled their lives. Their teachers attempted to teach them in the hope that they will be able to prepare themselves for college and jobs beyond their elementary schooling, but they believed that the environment may already have taken its toll on many of the children.

With the growing violence in America's cities, towns, and rural areas, young people are beginning to understand at an early age the meaning of mortality. Indeed, this situation presents a dilemma to today's teachers. Is the notion of making mathematics relevant naïve, given the influences with which many students must deal outside the school environment? What mathematics should they teach to their students? How do they get their students' minds off their problems beyond the school walls long enough to focus on developing their problem-solving skills? How do they get students excited enough in class to want to use manipulatives, graphing calculators, and other tools to learn mathematics? How do they get parents to help improve the educational experiences of their children? Are these endeavors even worthwhile, or should they be abandoned in favor of a back-to-basics approach?

The mathematical vision of the teachers at the Garnett School was not as clearly defined as those that documenters viewed at some of the other R^{3}M sites, perhaps because the school personnel devoted much of their time and energy to trying to improve the quality of their students' lives. Although these teachers faced challenges that differed from those of many of their cohorts at sites in the R^{3}M project, they were no less passionate or dedicated to their students and to their profession.

The teachers and administrators at the Garnett School were very concerned with their students' mathematics learning, but they were hampered by their lack of physical resources. They were unsure of how to make parents aware of mathematics education reform changes or of their desire to implement a new way of teaching and learning mathematics. The parents' lack of consistent goals and beliefs about what and how mathematics should be taught in the school were major stumbling blocks. Many parents believed that basic mathematics facts and skills should be stressed and that teaching problem-solving skills should be of secondary importance. The sentiment was expressed by two teachers:

> *At times I sense that we [teachers] are just functioning to keep our students safe during the day. You can look out the window and see a gang hanging out on the street corner or a drug deal happening right across the street from the school building. That's the reality we have to deal with. So it would be nice to do fancy things with Cuisenaire rods or pattern blocks or attribute blocks, but it's just not that easy. And then there are the parents. Parents want us to teach basic skills to their kids so they can survive in the world today. They want us to teach computation and more computation. They feel that is the best mathematics for their kids.*

I really work hard with my kids and want them to learn exciting things in my math class, but the parents want something different. A lot of the parents have to work. And they need day care for their kids. And the day care is not available. Or if it is available, it is not affordable. The parents of the population in our building can't afford it. They really can't. It's hard. For them, their first priority is getting their kids to master the basic skills. That's all. They don't want us to stress problem-solving skills. They just want their kids to be able to get jobs after school to help the family out. And that means working in a store and being able to give change or working in a restaurant and knowing how to make out the check.

Parents were concerned about their children's mathematical preparation. They truly wanted a good education for their children, and they wanted their children to be successful at mathematics. But for many of these parents, this meant a computationally based curriculum, not one filled with manipulatives and problem-solving activities. As one parent explained,

I don't let my son go out and play in the parking lot anymore. It's just too dangerous. I've seen too many kids gunned down for no reason. I feel bad for him, but I have got no choice. He's a good boy and I want him to be safe. So when my son goes out to play, I only let him play in the hallway of our apartment building. And even that's not always safe. Last week a drug-deal raid went down right in my own apartment building. So, for me I want him to get an education. A good education and that means learning how to read, write, and do arithmetic. Not that fancy new math stuff, just what he needs to know so he won't get [cheated] when he goes to buy something in a store. You know—adding, subtracting, and multiplying.

The teachers were realistic about their goals for their mathematics program. They knew that change would not be easy and not without its price, especially in one of the poorest schools in a very poor school district. They also knew that the issues were complex and not easily solved.

Through the interviews with teachers and parents at Garnett School, documenters became aware of a difference of philosophy about the teaching and learning of mathematics. The teachers wanted to stress problem solving and mathematical explorations, whereas the parents wanted their children to learn mathematics through rote exercises and practice. Parents were satisfied with a computational facility and wanted their children to remain in a drill-and-practice atmosphere.

The administrators were trying to cope with a poor, inner-city school and its associated problems on a day-to-day basis. Many of the teachers had a mathematical vision but found that the complex issues of dealing with the changing mathematics curriculum, the influence of the students' environment, and parental pressures were difficult obstacles.

SUMMARY AND CURRENT ISSUES

Repeatedly at the R^{3}M sites, when teachers were asked questions about what was new and different about their mathematics programs, they responded with some version of "now mathematics is more than computation." What is interesting about this mathematical vision is that it is not an articulate description of what mathematics in the grades K–12 arena is, but rather it is an articulation of what mathematics is not. The by-product is a vision that is neither rich nor powerful.

The stories from the R^{3}M sites presented in this chapter highlight the challenges that mathematics educators face in defining the mathematical content of the reform. These challenges include deciding what mathematics is relevant and meaningful for diverse student populations and being able to unite different mathematical experiences and cultures into a curriculum that is influenced by many factors: students' attitudes, needs, interests, and understanding; the types of assessments that are being used; the recommendations from professional organizations; and outside pressures from parents and the community at large. Insights gained from the documenters' descriptions of the R^{3}M sites suggest that the challenges of putting into practice a well-articulated vision of mathematics content have the potential to be overwhelming for the individual teacher. Developing and implementing such a vision is complicated because as a community of mathematics educators we are only beginning to realize that questions about which mathematical ideas, concepts, and procedures should be incorporated are deep and different in the proposed restructuring.

The visits to the R^{3}M sites occured early in the process of change envisioned by the *Curriculum Standards.* Few resources existed that expanded on the vision of the original document, which may explain why at many sites the explicit content of the mathematical vision was not clearly defined or why sites had difficulty going beyond "mathematics is more than computation." Since our visits, many supplementary resources have been developed, including a set of major curiculum projects that are being piloted in classrooms across the county and are now being distributed by commercial publishing companies. In addition, the National Council of Teachers of Mathematics has established a process to revise the *Standards* documents on the basis of the experiences to date. The process will be interesting to watch as it unfolds. Questions remain about what influence, if any, these projects and recommendations will have on efforts to change mathematics curricula and instruction.

Chapter 3

WHAT SUPPORTS TEACHERS AS THEY CHANGE THEIR PRACTICE?

Patricia P. Tinto and Joanna O. Masingila

In one classroom, a kindergarten teacher was working on patterns. She had made paper snowmen. Some had hats; others had no hats. The teacher placed snowmen on a large floor chart and asked the children to guess the pattern which consisted of a hatless snowman followed by a hatted snowman. Several children struggled with the pattern and continually responded with "snowman" when asked what would be next in the pattern. The students then started making their own patterns by pasting snowmen, with and without hats, on paper strips. While the students were busy with this activity, the teacher "interviewed" each of the students who she felt wasn't "getting it." Through these interviews, the teacher was able to assess that the students didn't realize that the hats were relevant. Being a snowman was the more important characteristic.

—R³M documenter

In the class just described, the teacher took the time to interview students because she was very concerned about what the children were thinking and about how they learned mathematics. She noted that in the future, she would have to discuss with these children the different pieces in a pattern. She believed that it was important that teachers be willing to learn as well as to teach. She remarked, "It's another indication [that] what you're doing and what the kids are seeing is completely different."

This teacher represents the kind of teaching we saw at Garnett Elementary School: a genuine caring about students and student learning, the ability to talk about what should happen in the classroom, the selection of appropriate resources and materials, the thoughtful preparation of mathematics teaching, and the willingness to reflect on teaching.

But even in this school, these teachers believe that much more can be done to help them and other teachers change the mathematical learning environment for every child at Garnett. For them, the change process is ongoing, one that requires not only dedicated teachers but also a range of supports to help all teachers change and improve their practice.

This chapter presents an overview of what we learned from the R³M sites about the types of supports that need be in place to foster the change process and about the ways that teachers made sense of those supports. These supports can be described as follows: gaining a professional role, ensuring staff development experiences, involving all teachers, risk taking and collaboration, experiencing institutional commitment, building community support, and having visible outcomes.

GAINING A PROFESSIONAL ROLE

Teachers who talked about successfully changing the mathematical learning environment for students also talked about having a voice in the process and being treated as professionals by administrators, peers, and community members. Two issues stood out. First, they saw themselves as being able to use their personal experiences as springboards for making changes. They noted that in the process of changing classroom practice, they were recognized for their role in making decisions that directly affected themselves and their students. Second, they became teacher-leaders in their schools and were encouraged to develop networks of professional relationships within and beyond the school. In several sites, these teachers became responsible for educating their fellow teachers and building wider networks for change.

Developing Mathematical Self-Confidence

At Southland Elementary School, teachers minimized the risks they felt in developing meaningful mathematics lessons by connecting mathematics to real-world situations in which they had expertise. Several teachers noted that sometimes change is easiest when we start with something we know and enjoy. The teachers at Southland Elementary School used personal interests as their vehicle for connecting students with mathematics. These teachers developed units, funded through local and state minigrants, that gave their students real-life experiences with mathematics.

Focusing on real-life contexts for mathematics meant that the school's current textbooks could not be used as the main source of mathematics learning. It also meant that students might need to use mathematics not currently taught at a given grade level. Some teachers found that they had to work with a particular area of mathematics that they believed was not their strongest. With the support of their principal, a mathematics specialist, the county mathematics coordinator, and parents, the teachers were willing to take those risks. The following unit exemplifies the teachers' commitment to "real-world mathematics."

A first-grade teacher created a unit on cooking. She noted, "I wrote a grant called 'Books for Cooks' and I just got a whole lot of different big books that had something to do with cooking and then we just do math with it. Right now we are doing *The Gingerbread Boy*."

The teacher explained that she used the big book for the class reading lesson and then introduced measurement and estimation through the cooking. Number sense was a focus in several of her cooking activities. She cited an example in which the children initially estimated that they needed 400 cups of sugar to make chocolate chip cookies and later refined their guesses to half a cup. "Guessing 400 cups of sugar to a little less than half a cup. So I think it's really helped with estimating." She went on to note,

> And we do some calculator activities, and I always bring the prices or the grocery list of what I bought for that particular activity. Even though right now they don't know how to add two-digit numbers and things, they can do it on a calculator; and so I think that they're learning how to use the calculator.

Developing Professional Networks

At Garnett Elementary School, three teachers were involved in a National Science Foundation–funded project. The intent of the project was to get the teachers to think about how students learn and students think and how they as teachers can learn along with the children. The project brought the teachers to a point of leadership in their school where they would then duplicate the project for their coteachers.

At another R^{3}M site, the Mathematics Teaching Project (MTP) started in summer 1991 as an in-service course for twenty-eight teachers from just a few schools in two of the city's poorest school districts. Organized by the districts' mathematics supervisors, together with a university professor, the course was designed to present the *Curriculum Standards* to practitioners. The ultimate course goal was for teachers from grades 4 through 8 to understand how to help all students succeed in algebra by grade 9. The assumption was that success in algebra would be more likely if students had appropriate grounding throughout elementary and middle school. As the MTP evolved, it expanded to offer mathematics summer classes to teachers from grades K–12 in all the schools in the original two districts. In addition, during the school year project participants organized their own conference—open to teachers from any of the fifteen school districts in the city—at which they shared what they learned in the summer course and showed how they apply this knowledge in their classrooms. The spirit of collaboration and sharing communicated by the MTP is perhaps best summed up by two teachers from Bedford Middle School who expressed the same sentiments in slightly different words. One said,

> *I think that one of the major things that the Math Teaching Project did was the camaraderie that it produced among all teachers, in other districts, and in your own district, and in your school. You have a team, you have support, and you have people to go to.*

The other teacher expressed the project's spirit of sharing in this way:

> *You put yourself forward as someone who wants to improve the math, and when you find kindred spirits, that's really nice, no matter where. Whether it's the same district, different school, same school.*

The MTP teachers believed not only that their teaching was different as a result of their work with the project but also that their stance toward their profession was different. Even while admitting that they still had a lot to learn, they described with confidence the decisions they made each day in their classrooms. They looked forward to attending workshops and conferences, where they often found themselves in leadership roles. In essence, they were developing a network of colleagues with whom they shared interests and expertise. Fellow teachers came to them for advice, and they, in turn, consulted with others when they had questions or problems. Newcomers were encouraged to become part of the network.

ENSURING STAFF DEVELOPMENT EXPERIENCES

Teachers who talked about successfully changing their teaching and developing more effective mathematical learning experiences for children also talked about having ongoing, effective staff development experiences. These experiences were defined as those that challenged teachers' beliefs not only about what mathematics is but also about how children learn mathematics.

Deep Brook Elementary School's story documents the importance of staff development for meaningful change in the classroom:

> *There is noise in the classrooms. It is the sound of students actively participating in explorations and investigations which will make them mathematically empowered. Some students are collaborating on mathematical ideas, devising their own algorithms, or writing about mathematical concepts, while others are using manipulatives, calculators, computers, or pencil and paper as a means of aiding their explorations. Listening to students hypothesizing, investigating, validating, and justifying their conclusions to their*

classmates are a few of the observations one makes during a visit to the mathematics classes at the Deep Brook Elementary School.

—R³M documenter

The principal and superintendent offered their insights on the changes that these teachers were able to make in their mathematics instruction:

The single most effective and most efficient—cost efficient and time efficient—way of helping teachers make change is to provide them with staff development workshops. In some instances there are central office administrators who really won't take the time to understand the dynamics of staff-development workshops and the long-range value to the children of the district and the positive way in which teacher attitudes toward mathematics can be affected.

—Principal
Deep Brook Elementary School

Staff development is very important. You are asking teachers to teach in a different way. You're asking teachers to organize their classrooms in a different way. Classroom groups being noisier than they might have been. Everybody moving pennies or blocks or chips around is very different from students writing in a workbook. So whenever you go through those kinds of changes, you must invest time and probably some money in training. So take care to build the staff development program.

—Superintendent
Deep Brook School District

The mathematics education reform efforts that were under way in the Deep Brook School District were the result of an administration that supported allocating significant funds for staff development activities. Rather than have teachers participate in a prescriptive process, the district advocated a developmental learning approach for its staff, who were treated as thoughtful practitioners and were encouraged to view themselves as learners. They became engaged in a process whereby they were asked to investigate and to construct knowledge, just as they would ask their students to investigate and to construct knowledge. The principal noted that the district had a steady growth in attendance at workshops. As more teachers began to participate in these workshops and other in-service opportunities, other staff members reported that they began to notice their enthusiasm. As time progressed, they began to communicate about the dynamic lessons, challenging activities, and different approaches to mathematics teaching that they had learned from the workshops. These conversations created an interest in other teachers who eventually enrolled in staff development workshops.

In attempting to describe the meaningful experiences provided by the mathematics in-service programs, one teacher noted,

> *When I first started taking the workshops I started seeing math in a different way. I started understanding things that I had only memorized before. If at my age I'm just realizing these things, wouldn't it be nice if children could see that from the beginning? If they could realize that there are lots of different ways to solve a problem and how to verbalize the way to solve a problem.*

A kindergarten teacher added,

> *When I first started teaching, I initiated the questions of what I was trying to get the children to think about for that day. I was pulling what I wanted out of them by asking them specific questions. As I've gone through the years, I've become less of the center, so I don't ask specific questions anymore. Instead, I really try to make it their lesson now and let them tell me what they need to know.*

Several teachers described how the workshops had given them a more encompassing view of their own mathematics teaching. One third-grade teacher explained, "Before taking the workshop, I would focus my math classes on computation skills. Now, after attending the workshops, my focus has expanded greatly to include a more problem-solving approach. It still includes a computational-skills component, it's just not the main focus anymore." Statements like these, expressed by the majority of the Deep Brook teachers, prompted the administration to continue allocating significant amounts of the district's funding to staff development in mathematics.

INVOLVING ALL TEACHERS

Teachers who talked about successful change in developing mathematically powerful students also talked about involving all colleagues, administrators, and community members on a schoolwide and districtwide basis. Teachers noted that the capacity of individual teachers to effect a change in their classrooms and provide powerful models for other teachers to follow implied a commitment beyond a single classroom. It meant involving teachers from many quarters of the school and school district.

Inviting the Involvement of Every Teacher

As a result of eight years of ongoing, effective staff development practice, the entire staff of Deep Brook Elementary School recognized the importance of giving students a solid foundation in mathematics. All twenty-one classroom teachers and most of the special subject-area teachers participated in mathematics in-service workshops. Even the art, music, and physical education teachers, along with the library media specialist, used materials developed in the workshops to enrich aspects of their own curricula that related to mathematics. The combined efforts of the staff gave an observer a very broad sense of a well-planned and integrated learning experience for the students.

Developing Districtwide Involvement

In the Parker Springs School District, twenty teachers from around the district are accepted as grades K–3 mathematics specialists in each year as part of an externally funded project. At the time of the R^3M project, there were twenty first-year mathematics specialists, twenty second-year mathematics specialists, twenty third-year mathematics specialists, and nine fifth-year mathematics specialists. There were no fourth-year mathematics specialists because that was the year between the pilot and the actual funding of the project.

The essence of the program's underlying strategy was to rely on teacher change in the primary grades as the mechanism for bringing about change in the entire district's mathematics program. We were told that in the context of Parker Springs, at least two arguments for relying on this strategy could be found:

> *Though the district has developed a Core Curriculum guide based on the NCTM Standards, the district's site-based management program has meant that each school has the final say on curriculum.*
>
> —R^3M documenter

> *The district administration had been impressed by the extent of teacher change effected by a writing-process initiative in the primary grades and by the consequent effect on children's writing. The superintendent recalled listening to five-year-olds "critiquing each other and talking about style and form," sounding "like a group of college people sitting around." According to the superintendent, the values of the writing program transferred to the mathematics program.*
>
> —R^3M documenter

Engaging Teachers through District Goals and Curriculum Development

In the Green Hills School District (GHSD), gathering ideas from all teachers helped promote teacher ownership of newly written curriculum guidelines. Using the district's philosophy, goals, and objectives, teachers wrote suggestions and activities for implementation. These were classified by objective and type of activity and were included in the curriculum guide. Teachers needing ideas for a particular objective could begin with the guide itself. Suggested instructional material, available in the district, and links to technology were included with each activity.

The GHSD 1992 curriculum document was the product of various planning committees. In the foreword of this document, the mathematics supervisor stated the following:

> Never before has such an impressive and extensive process been employed to revise a math curriculum in our district. Never before has a math curriculum revision involved such camaraderie, teamwork, and agreement from such a large number of teachers, patrons, and administrators. Never before has the professionalism, dedication, and effort poured into a new math curriculum been more appreciated. Every [Green Hills] teacher contributed to this new curriculum. You all deserve a vote of thanks.

After the research and writing committees completed their work with the curriculum guide, the other committees began reviewing instructional materials, writing a handbook for parents, and planning in-service opportunities for teachers and parents.

RISK TAKING AND COLLABORATION

Teachers who talked about successfully changing their teaching also talked about collaborating with other professionals and organizations. They valued having colleagues with whom to share their successes and their risk taking. They valued having opportunities for teacher-to-teacher dialogue and saw the need to expand that dialogue by inviting more teachers into that conversation.

At Desert View High School, teachers who early on committed to developing the curriculum-projects approach still maintained a patient and flexible stance with regard to the other teachers—acknowledging their colleagues' early fears, openly sharing their own successes and failures, and creating channels of support for others to take risks *when they were ready to do so.* Consistent with this image was a district administrator's characterization of these teachers who "are moving in the direction appropriate to the program":

> *They are asking themselves a lot of questions about what they are doing and they are trying things they never did before—taking risks in the classrooms in terms of the way they teach.... [T]hey have come to realize that everything they do doesn't have to succeed. They have talked about their successes and failures. When one [teacher] has not been as successful as the other, that*

successful teacher has said, "Well, I did it this way"—giving the second teacher a chance to think, "Well, I might try that again and tinker with it a little."

A comfort with risk taking and the freedom to fail cannot be mandated, but they can be nurtured. At Desert View, a policy of equitable teaching loads, a sense of professionalism enhanced through support and respect from university mathematicians, and the provision of time and resources for ongoing collaboration with colleagues enabled teachers to come to terms with a mathematical vision in a deliberate and nonthreatening manner. Desert View's experience with project-driven reform demonstrated well the value in two-way exchanges of innovative ideas between public school practitioners and university mathematicians. Perhaps the most significant and replicable aspects of Desert View's efforts were the opportunities created for risk taking and long-term collaborative planning among teachers. Moreover, it is clear that such opportunities were optimally fostered within a climate that valued teacher ownership and autonomy.

Four meetings a year were instituted in which university mathematicians and high school teachers could share concerns, strategies, and long-term plans. University mathematicians provided feedback on the mathematical content and quality of teacher-designed projects, although Desert View's teachers quickly assumed ownership of the design and modification of materials to meet the needs of their students in the high school. One Desert View teacher remarked,

Certainly the feedback and interest we have received from the principal investigators [mathematics professors] on the project program has made a big difference. Because we have somebody outside the system who has been interested in what's going on, who has been in communication with us, who has asked provocative questions, we have more pride in what we do. We think of ourselves more as professional educators and not as just holding down the fort day to day.... The principal investigators have always afforded us respect for what we do at our level.

Another Desert View teacher echoed a similar perception: "I feel that they [the university people] don't tell us what we have to do; they just offer ideas to us and we adapt things the way we want—and that's OK."

These perceptions matched those expressed by the university professors who served as principal investigators in the program. When asked about the respective roles envisioned for the university and high school personnel, one mathematics professor responded:

We needed the high school people to help us with the logistical part of doing with high school students what we had been doing with college students.... We offered the pedagogy and the mathematical

strength and they offered teaching experience and understanding of the society involved. The original proposal was very egalitarian—we have expertise in some stuff and you have expertise in other stuff, and if we put them together maybe we can get a good system going.

In part because of the mediation role played by university mathematicians and the structure provided by the quarterly meetings and in part because of the long history of camaraderie among the faculty, a new mindset of openness and sharing was fostered. Teachers appeared eager for opportunities to communicate mathematics to one another as a means of strengthening their own conceptual and methodological understandings. One local administrator described the value of this transformation, noting that the project presented teachers with an environment in which they "have a chance to try new ideas and talk to each other about their successes and failures—a whole new kind of peer coaching for us." Nearly all teacher interviews indicated a "closeness" among mathematics teachers. One of the cochairs indicated the following:

I think the biggest thing that has happened with this [collaboration] at the university is it has opened our eyes to the possibilities that are there and made us feel free to try and free to fail because we have all had successes and failures, and in this department we have such a good camaraderie that if you do something and it just completely flops, you are not embarrassed to go tell the guy next door.... Our math department staff has never been closer. We have meetings once a month for an hour or two after school about our projects, how they are going, and give each other hints and suggestions and things. We are not competing at all.

As an assistant principal at Desert View observed:

One thing I've learned from this experience [with project-driven reform] is that you can take a fairly traditional high school in a very traditional setting and you can break some ground, you can do some nontraditional things that have not been imposed by edict or force. No one said that this had to happen.

EXPERIENCING INSTITUTIONAL COMMITMENT

Teachers who talked about successfully changing their teaching practice talked about support in terms of school-based commitment. They noted the importance not only of having ongoing positive relationships with administrators and fellow teachers but also of knowing that these administrators and teachers were committed to the same goals and were willing to support them with concrete actions.

Common Planning Time

Like Desert View, the Deep Brook School District had a strong collaboration with a university program. However, teachers there also expressed that equally strong commitments must come at the school level. One teacher noted the prevailing mood:

> *What's nice is that we've all basically taken mathematics courses together. When things come up that are good for what we know we need to be teaching at that grade level, you can share and say, "I have a really great idea, you might want to try it in your room."*

Several teachers emphasized the link between staff development and the opportunity for teachers to have common planning time at least three times a week. At Deep Brook, teachers of the same grade were given a block of planning time once a week. They cited this meeting time as a valuable asset in their sharing and teaching:

> *Teachers can brainstorm ideas together and share activities from journals and materials from conferences. We plan activities together and devise common themes to use with them, such as graphing data from pumpkins and apples, crayfish, and dinosaur measurements.*

Teachers were required to commit to at least trying new strategies learned from workshops—this was part of the principal's concern for follow-through and meaningful in-service training. The principal, in turn, routinely volunteered to model appropriate mathematics instruction at all grade levels within the school. One teacher surmised,

> *I think you need lots of fellow teachers around you to want to change also. You need the support and you need the meetings and you need the development from the district. So I think that staff development is absolutely necessary, and that has to be districtwide. And I think you have to have a buddy, one or two buddies that want to do it along with you. And you have to get together and you have to meet once a week and you have to sit down and hash through it and that's taken me seven years to get my program to where I want it to be and it's still not there. Next year it's going to be something else.*

After-School Working Dinners

The Queensborough School District comprised two schools: a grades K–6 elementary school and a grades 7–12 junior-senior high school. Communication and collaboration between teachers from the schools was possible because the two buildings were adjacent. Cooperation and collaboration among teachers at all three levels—elementary school, middle school, and high school—were developed and facilitated through evening meetings.

Once a month a consultant visited the district to meet and work with teachers during the school day. During the visits, the consultant spent full days at both schools, answering teachers' questions and providing demonstration lessons during which teachers were released from their own classes to observe. These lessons were also videotaped. After the school day, the consultant, the superintendent, the principals of the elementary school and the high school, and all the mathematics teachers of grades K–12 met to investigate mathematical topics. At each evening meeting, teachers presented lessons on a selected mathematics topic. Sometimes these discussions included collective reflection on the videotaped segments.

The superintendent used district funds to give teachers a pizza supper so that the meetings could extend into the evening. He believed that the evening sessions greatly influenced the participants' views of mathematics and teaching. A high school teacher agreed:

> *The meetings are important because to understand how children learn mathematics is really relevant to all age levels, and that's something that all of us benefited from--and how to bring some concrete manipulative activities into the classroom. That [the evening activity] was more geared toward elementary, but nevertheless some ideas were sparked for high school teachers as well. It was important for us to see what's happening at the elementary level and what the kids are coming through when they come to us. So even that alone was really beneficial. We really come away with an awful lot.*

The Freedom to Experiment

At Manzanita High School, Mr. Bell, a mathematics teacher, described the district mathematics program and his mathematics teaching in the following terms:

> *The whole district appears to be on-line with the Standards. I feel like we're experimenting ... and it takes guts to experiment because it kind of puts you out of your comfort level. But it also takes guts on the part of the district to back us.... We've really been given a lot of support from the district giving us the freedom to experiment as well as the financial support to purchase all of the manipulatives we need.*

Specifically, Mr. Bell characterized his mathematics teaching in terms of problem solving, relevance, the use of manipulatives, group work, and the promotion of communication. He characterized his sense of mathematics reform as the "freedom to experiment." Teachers were continually reflecting on their practice, gathering data, interpreting the data, reinterpreting their practice, expanding their repertoire, and trying new approaches weighed against old "proven" practices. At Manzanita, change has become accepted as the norm, not the exception.

BUILDING COMMUNITY SUPPORT

Teachers who talked about successfully changing their teaching practice also talked about the importance of community support not only for classroom-based changes but for staff development as well. In the Deep Brook School District, perhaps one of the most difficult ordeals for the administration was giving parents quality assurances regarding the education of their children. The administration launched an awareness program for the community by organizing evening meetings for parents based on the Family Math program, having the principal give presentations on the *Curriculum Standards* at open house meetings, and having teachers communicate information about student progress on a regular basis.

In the Oldburg School District, educators were also working to gain parent support for the changes in their program. The mathematics specialist explained a common problem:

> *One of our biggest problems in the beginning was that kids would go home and say that they hadn't done math and it had been for weeks ... so it finally got to a point where third-grade teachers on days that they used counting blocks or did puzzles, they would put at the top that this was your math today or they would say to the kids as math class was going on, "This is math class today; this is what you are doing in math."*

Students were helped to explain what they did in mathematics class to their parents. After working on a lesson involving words such as *arrays* and *tessellates,* students were given a worksheet in which they were asked to make different designs with an area of twelve color tiles.

The mathematics specialist told the fourth-grade students, "Underneath [your designs], write Area = 12 squares so that when the paper goes home, you can explain to your mom and dad what you did. If you want to write it like mathematicians, write it, put A = 12 square units." The specialist commented that in hindsight, the district wished that it had been proactive in educating parents. Instead, it had to be reactive and tried to educate parents through such programs as the PTA.

COMMON THREADS IN SUPPORTING CHANGE

Not every school district or even each school observed had in place all the support structures for reform. However, those sites where teachers spoke with confidence about reform efforts, where students appeared actively engaged in meaningful mathematical experiences, and where administrators and parents were also able to participate in the dialogue about mathematical reform were also those sites where it was evident that several of the support structures were in place.

East Collins, Desert View, Deep Brook, and Green Hills gave vivid images of mathematically powerful students and professionally empowered teachers. Each site not only had most of the support structures, briefly discussed above, in place but also had taken the next steps in the planning stages. Issues of assessment at the student, classroom, and school level were beginning to be addressed. East Collins was also discussing the issues connected with taking a school-based reform effort to other schools in the district. Both East Collins and Green Hills had key people from the early years of reform efforts leave, and both sites noted that although these people were missed, the process of reform was so much a part of the school culture that the programs continued. When the teachers at East Collins talked about sustaining reform efforts, it became very clear that they were talking about reform at the classroom level. They were very much aware of one another's efforts and areas of expertise. They were proud of their colleagues' attempts at trying new ideas. To get past the traditional mindset of teaching and the regularities and routines of classroom life, these teachers had opened their classroom doors.

The data from all sites where teachers agreed that classroom practice was fostering improved mathematical learning can be summarized by the following statements:

- Ongoing discussion among teachers about classroom practice takes place, and teachers take responsibility to sustain that conversation.
- Formal collaborative arrangements among teachers, administrators, business partners, university consultants, and community people foster a heightened sense of professionalism among teachers and help develop an atmosphere of trust in which teachers' input is valued.
- Ownership of reform efforts is strongest when the focus of the reform effort evolves from all stakeholders' concerns and interests.
- Leadership is necessary throughout the reform process, but its presence must take on varied intensities and have support from all teachers involved.
- Funding is supportive when it can be accessed by the teachers directly and used for purposes deemed useful by those receiving the funding.
- Events and people that are seen by teachers as supporting reform efforts were those that help teachers construct the knowledge they need to be professional educators.

CONCLUDING COMMENT

In the context of this chapter and your own school, reflect on the following questions: How do communities of learners develop? How can school administrators provide the level of support needed for these communities to thrive and for the level of collaboration among teachers to increase?

One feature common to all the sites in this study is the importance of the individual classroom teacher. Every effort at reform must come to fruition in the classroom. The individual teacher is the hub for all reform efforts and the lens through which change in classroom practice is viewed. It is also true that teacher change is fostered by a range of supports.

Chapter 4

WHAT LEADERSHIP MODELS ARE SUCCESSFUL IN REFORM EFFORTS?

Joanna O. Masingila and Patricia P. Tinto

What we find in the instructional supervisor is somebody who challenges all the time. If he says we're going to do it, we're going to do it.

—Mathematics teacher
Scottsville High School

It seems like a lot of times a good idea comes up and everybody gets excited, and for two or three years it is there. Then it just kind of fades away. And I [am] really feeling strongly about this. I don't want this [program] to do that. And kind kind of selfishly, I was afraid that if I didn't accept being head of the math department, that somebody else who wasn't interested in our changes would let it fizzle, and I don't want that to happen. So I'm hoping that I can continue to push it.

—Mathematics cochair
East Collins High School

When we consider how to reform the mathematics teaching and learning at a school, questions about leadership arise. How can we get started? What form should leadership take? Who will lead us in the reform? In all the schools and districts in the R^3M study, identifiable leadership for the mathematics education reform agenda was a central factor in its success. Leadership can be thought of as the ability to move others in a particular direction. In the schools and districts we visited where movement was taking place, leadership was evident, but it appeared in different forms in different places. In some situations, one person was the driving force behind reform in mathematics teaching and learning; in other places, leadership came in the form of a districtwide or schoolwide committee, with a wide representation of people. In still other situations, a group of mathematics teachers led the way. In this chapter, we examine these different leadership styles and analyze what made them work.

CAN REFORM BE LED BY A SINGLE PERSON?

One model of leadership that we observed was that of a person who had a vision for changing the way mathematics is taught and learned in his or her school or district and who then worked to see that vision implemented. In each situation where we observed the single-person leadership model, this leadership for reform efforts came from people at the administrative level. For example, the teachers at Green Hills contributed to the development of the grades K–12 mathematics curriculum, but it was the newly appointed district mathematics coordinator who organized and led that effort. For Pinewood High School, it was the county mathematics supervisor who had the vision for mathematics reform that would have an impact on the school. According to one of the mathematics teachers:

> *We really do have a dynamic supervisor in math who is very energetic and really pushing for change and just really active.... She's well respected, and I think everybody respects her and appreciates what she's trying to do.*

The mathematics supervisor was also the leader in change at Scottsville High School. Teachers and administrators recognized his ability to get what he thought was needed for the mathematics department and his skills at motivating teachers to try new things in the classroom. According to the school principal:

> *I also think the reason the math department is so willing to try new ideas and do the kinds of research that they themselves were, is primarily the supervisor's leadership. I mean, he will not allow them to sit back on their laurels. I mean, if they do something really good and really unique, all of a sudden he wants them to do something else. They don't sit back and enjoy the fruits of their labor that long. I mean, they're just moving on to something else.*

We next examine how the single-leader model was evident at four R^3M sites.

Deep Brook Elementary School

Deep Brook Elementary School is one of three elementary schools in a suburban district of approximately 2 000 students. The majority of teachers we interviewed at Deep Brook described the motivational role of the principal. In the words of a third-grade teacher:

I think, number one, you need an administrator who understands—not just paraphrases but really understands—what mathematics is and how children learn mathematics. Then you need a district that's committed to feeling that the training and support of teachers are paramount.

The pivotal role of the principal in this setting cannot be overemphasized. He was the major catalyst, encourager, and reinforcer of change and advances not only in the Deep Brook School but throughout the district. He had been the driving force behind districtwide mathematics reform for eight to ten years. He was the means through which the teachers became aware of the *Curriculum Standards.* We found that the principal himself led workshops on mathematics programs and had served as a speaker at local, regional, and national mathematics conferences and as a resource for schools throughout the region. The central role played by the principal was recognized by the superintendent and by every teacher with whom we spoke:

Clearly the most significant factor has been the building principals who have become knowledgeable in the areas of mathematics and mathematics instruction—who have become knowledgeable in staff development and adult learning processes, who have implemented a staff-development program, and who have communicated a vision of mathematics on a long-term basis to the staff and to the constituents in the schools.

—Superintendent
Deep Brook School District

The principal at Deep Brook expressed to us his belief that teachers should assume a leadership role in any change that occurs in the programs within the school. He saw himself as an agent of change, but he also realized that for any change in instruction to be effective, it must be initiated by the teachers, and students, themselves. Thus, he advocated transfering the program from a prescribed curriculum adopted by the district board of education to one owned by the teachers and the students.

We found that teachers in the building complimented the principal for his dissemination efforts that included providing research and other professional writings in his interactions with them individually and collectively. He often placed copies of articles and documents in their mailboxes. These readings appeared to be helpful because the teachers frequently referred to them as they collaborated on the redesign of the curriculum. The principal also made sure that upcoming workshop opportunities, conferences, and new ideas and themes in mathematics education were announced at faculty meetings.

Parker Springs School

Documenters visited an elementary school located in a large and expanding urban area and within the Parker Springs School District, which is among the twenty largest districts in the United States. In the late 1980s and early 1990s, the school district applied for and received a series of grants to employ grades K–3 mathematics specialists in each school. The intent of the project was to supply in-depth training and support for in-service teachers from each of the eighty-five elementary schools; these teachers would then serve as agents of mathematical reform. The force behind the project was the program consultant for elementary school mathematics in the district. She and her assistant offered ongoing support for the mathematics specialists through demonstration lessons and classroom observations.

At the beginning, the program consultant reported being the only person interested in applying for the external funding. She vigorously pursued the first grant, awarded in 1988, even though neither the superintendent nor the associate superintendent for elementary education was particularly enthusiastic about the program when it began and neither administrator did anything to encourage expansion. The associate superintendent characterized the project during the early years as "one of those satellite things that operates out there, and as long as it doesn't bother you and you don't bother it, you easily co-exist." Perceptions changed, however, as the associate superintendent noted:

> *We started to see the teachers beginning to be more enthusiastic. We saw a sense of professionalism begin to emerge and that they were meeting together. They were talking about concerns that they had. They were doing some problem solving, they were doing some rather forward thinking. They were attending the national NCTM [National Council of Teachers of Mathematics] conference, which gave them a sense of professionalism and pride that they hadn't had before. We also found that there were some changes taking place in those classrooms that caused other teachers who were near their rooms to say, "Hey, what's going on there? What are you doing that's causing this behavioral change in students?" Not only in terms of what they are learning in mathematics but just the whole attitude of students becoming more successful, feeling an enhanced self-esteem.*

Although it can happen that a program launched by a person in the central office soon slides into oblivion, in this instance the program consultant remained vigilant as the mathematics specialist project continued. She remained concerned about specialists' needs, worked to remove obstacles from their paths, and, in the words of one principal, actively "promotes the program and keeps it visible" among principals.

Green Hills United School District

Although many teachers contributed to the development of a grades K–12 mathematics curriculum in the Green Hills United School District, the newly appointed district mathematics coordinator actually organized and led that effort. Her enthusiasm and sense of purpose were strongly felt throughout the district, a midwestern suburban district of approximately 11 000 students.

Many teachers and administrators agreed that it was the coordinator who had the vision, as one teacher noted, "to see where we could go and get the rest of us on board." When we asked the coordinator how this curriculum project differed from others done in the past, she was quick to point out that this project was an effort to reform the teaching and learning of mathematics in the district. She stated that two important events helped shape the successful outcome of the project. First, both the *Curriculum Standards* and the state guidelines were available as guides to help direct the mathematical vision of the district. Second, the project occurred as the district was moving to site-based management and trying to involve all stakeholders—teachers, students, administrators, parents, community people, business leaders—in the process of decision making within the district. Thus the development of the mathematics curriculum was not left to the district mathematics coordinator or to a few teachers, as in the past, but involved all stakeholders. With a variety of people involved in the curriculum development process, the district mathematics coordinator had a very difficult job but was able to facilitate the workings of these different stakeholders by involving them on crucial decision-making committees.

Scottsville High School

Scottsville High School is a high school with approximately 2 000 students, situated in a suburban area. From our observations at the school, and interviews with all the mathematics teachers and administrators, the instructional supervisor for the mathematics department emerged as a significant force in the innovation effort at the school.

The instructional supervisor has provided the vision for the mathematics department for ten years and has helped bring about technological and curricular reform. He believed that one way to innovate was to move the department into technology and that the best way to accomplish this goal was to get a computer laboratory. One teacher noted, "He really pushes. Everything that we have, he's pushed for." Another added, "And pushed and pushed and got it. And once he got it, he's tried to negotiate for us in terms of time because he very much wants to have some of this time. So he's 100 percent behind us in terms of trying to find time for us to work with the computers. He encourages us to do as much as possible in the lab." A third teacher stated, "He said five years ago we would be at this point and we all looked at him and said, 'Yeah, right.' And we are at that point."

A fairly new teacher commented on the atmosphere created by the instructional supervisor: "There is an atmosphere, and you see everybody around you hustling and bustling and getting a calculator, sitting at the computer and working with students at their desks, and you just—even as a new person—you just get sucked into it ... the whirlpool is there." All the teachers and administrators at Scottsville High School agreed that the instructional supervisor was central to the changes made so far. When asked what other schools could do to reform mathematics education at their schools, one teacher commented, "You need someone strong ... at times ... we overlook the fact that [the supervisor] has gotten us to where we are today."

At each of the R^3M sites discussed in this section, the single-person leadership model came from people at the administrative level. What are the advantages and disadvantages of such a model? Can reform be sustained under the single-leader model if the leader leaves the school? What types of supports would be necessary for the reform efforts to continue? What type of environment would be necessary to have the single leader be a teacher? What are some common threads among the sites where change efforts were led by a single person?

REFORM LED BY A GROUP OF PEOPLE

A group of people supplying leadership for other people in a school or district is a second model of leadership that we observed. This model varied among the sites. In some instances the leadership group involved teachers and administrators; in other places it involved university professors along with teachers. For example, reform at East Collins High School was led by two teachers (one of whom was the mathematics department chair), the district mathematics coordinator, and an associate principal, whereas the entire mathematics department and a group of mathematicians at a local university became the leaders for Desert View.

The chair of the mathematics department at East Collins accepted a leadership role in the planning and implementation phases of the mathematics program because he was afraid that left in the hands of others, the effort would not succeed.

> *It seems like a lot of times a good idea comes up and everybody gets excited, and for two or three years it is there. Then it just kind of fades away. And I really feel strongly about this. I don't want this [program] to do that. And kind of selfishly, I was afraid that if I didn't accept being head of the math department, that*

somebody else who wasn't interested in our changes would let it fizzle, and I don't want that to happen. So I'm hoping that I can continue to push it.

The other teachers at East Collins believed that they were an important part of the decision-making process as their mathematics program unfolded. According to one teacher, "They didn't make a move that they didn't consult with all of us."

After the one-year start-up period in which three teachers implemented projects in their mathematics classes at Desert View, the remaining teachers in the mathematics department became participants. Interestingly, the three who started the program did not continue on in any strong leadership capacity. All participants seemed to share equally in developing new projects and moving the program forward. According to one of the teachers:

When we first got together as a project group ... there were definitely some barriers between us as far as being afraid to criticize or damage egos. Through the process of editing projects for each other, we have taken each other's worth. And it's helped level everybody out to an even playing field where they feel more comfortable communicating back and forth.... Getting over the barrier of having done something dramatically different in the classroom, and being successful at it, has made everyone believe that it's not that tough to try new things. Now it's almost expected that you are going to try new things, see what happens, and not be afraid to report back to your peers about what worked or what didn't work.

We next examine in more detail how the model of shared leadership was evident at these two R^3M sites.

East Collins High School

East Collins High School, located in a southern state, is one of two high schools in a suburban-rural district of approximately 12 000 students. Reform in mathematics education at East Collins had its impetus in a partnership with industry. The Perfect Situation for Mathematics Learning (PSML) project came to life as a partnership among several important constituencies (see p. 21). Teachers recognized that a large number of East Collins students were choosing to enroll in nonalgebra mathematics courses in their ninth-grade year. These students were not being challenged mathematically. The associate principal and the district mathematics supervisor also recognized the need to have more students take algebra and stressed the need to bring relevant mathematics to the district's students. Additionally, recent state requirements mandated that all students graduating from high school must have completed algebra. However, recognizing a need and implementing change are very different processes within public schools. The catalyst for change came in the form of a business partnership. The mathematics chair explained how power in the original committee shifted as the implementation phase began its early stages:

> *After we started last year, the committee voted to make our associate principal the head of the committee, and the money was put into an account in our school system. And I handled the disbursement with her approval. So we, the two lead teachers in the project, the associate principal, the county mathematics coordinator, now pretty much make the decisions. It's up and running and the other folks have said, "Well, it's yours now." And they're still there, and we still meet, and I like to hear what they have to say and they have real good ideas. But basically we're making the decisions.*

Even though the PSML project had an identifiable group of leaders, we observed that all the mathematics teachers felt ownership of the project. In fact, we observed that the reform in mathematics teaching and learning that was under way at East Collins High School was successful, in large part, because of the way that authority over, and ownership of, the project was shared through a number of two-way connections. At the core of all these connections were the mathematics teachers at East Collins. The teachers have shared-authority relationships with business representatives, with school administrators, with one another, and with their students.

The business partnership that was the catalyst for creating PSML started when the state power authority and a large regional industry wanted to buy for the school a computer-based algebra program that could also serve adults in the community through evening classes. Several teachers and administrators visited a school that used the computer-based algebra program, but they decided that it did not fit their philosophy. Even though the school did not adopt the computer-based program, the business partners were still interested in working with the school, which brought about the creation of the reform project. From the reports we heard, the teachers appeared to have played a major role in formulating the philosophy and in deciding how the reform would be implemented. One teacher spoke of the empowerment gained by the teachers when they were able to design what they taught and how they taught: "We were able to use what we knew was correct and what we wanted to do."

Teachers noted the support they received from administrators at the school and district level in implementing the PSML project. An important endorsement came when the teachers requested that the mathematics sequence be changed for the third year of the project. That is, they wanted geometry to follow first-year algebra, with second-year algebra taking place in a third year. The original sequence had been first-year algebra, second-year algebra, and then geometry. The district approved this idea, which validated the teachers' authority over the curriculum, even though the change caused a shortage of recently adopted textbooks for the third year of the project. The associate principal went to the district office and obtained enough new geometry textbooks for all the classes at a time when textbook money was very tight around the state. As part of the arrangement, the other high school in the district would be given the extra textbooks in coming years. The ability to have some authority over their classes was seen as empowering by the teachers. The fact that the district provided additional funds was a sign of endorsement for the efforts of the teachers through the PSML project.

Although during the first year the project philosophy was tried out mainly by two teachers, more teachers came on board. By the third year, all eleven teachers reported that they endorsed and owned the PSML philosophy. One teacher explained:

> *I think one of the things that makes our math department so successful ... is that PSML isn't something that began last year here. It's been in the minds of these people—all of us—in one way or another for years and then it's just that PSML is the focus of that.*

Teachers reported to us that an important part of the PSML philosophy involved helping students become more autonomous learners. This change seemed to happen as teachers had students work in cooperative groups, asked questions rather than gave answers, and expected students to read material in their textbooks to learn new content. One teacher mentioned that the students began to see that the teacher was not the only authority in the room: "When they get to a question, they'll ask somebody else in the class before they'll ask me." Teachers noted that as students relied less on the teacher and more on their peers or on themselves, their confidence grew. One teacher observed that her students became "able to conquer math."

The crucial element in all the shared-authority relationships that existed through the PSML project appeared to be the central role played by the mathematics teachers at East Collins High School. The department chair summed it up when he noted, "That's why our faculty accepted it so readily—they knew that [the other teacher-leader] and I had major input from the beginning. As more of the control came to us, they saw that, and it wasn't like some outside group coming in and saying, 'Here it is.'"

Desert View High School

Desert View High School serves approximately 2 000 in an urban, southwestern district. In 1988, a group of mathematicians at the local state university, supported by a $225 000 grant, began to experiment with a calculus curriculum that involved the use of projects—extensive multilayered problems that students had to solve and then explain in technical reports. For the mathematicians at the university, the project approach represented not simply a new curriculum but a new mind-set for mathematics instruction. One of the program's prime university mathematicians described this as a shift toward greater student ownership of the learning process:

> *We gave [the students] a reason to become involved with mathematics. One of the biggest things we did was change the time frame in which students solve mathematics problems. It used to be they could solve five to twenty of them in an evening. Now it's one problem that takes a week or two—so you view the whole problem differently. You start out with something you absolutely can't do, and a little while later you end up with something that you can do.*

That same year, a small group of secondary school teachers in the local school district were completing their fourth year of monthly meetings as a mathematics curriculum committee. Originally formed in 1984 for the purpose of curriculum revision and textbook adoption, the committee members—a large proportion of whom were from Desert View—chose to maintain the group as an ongoing forum for the exchange of ideas among mathematics teachers in the district.

How to tailor instruction for students who had never experienced success in mathematics, on the one hand, and how to prepare academically motivated students for college, on the other, were primary concerns of Desert View's mathematics teachers. The documenters noted the following:

> *Even with these structures in place, early contacts between the university and the Desert View schools were more serendipitous than systematic. In the summer of 1990, complementary concerns brought together university mathematicians with teachers from the mathematics committee, in particular, several from Desert View High School. The university personnel were motivated by the desire to improve the preparation of students entering their program, particularly in terms of their ability to solve problems and write technical reports. More to the point, several of the mathematics professors sought an effective means of socializing students into the project approach. "Quite frankly," noted one professor, "we wanted to avoid the battle with students over whether they should write in mathematics class. We thought if they did this in high school, they would see it as a legitimate way to do work when they got to the university." One of Desert View's teachers suggested, more bluntly, "The university people wanted better projects without having to teach [the students] how to do them—but it clearly met our desire for instructional innovation as well."*

The mathematicians introduced the idea of using the project approach to teach high school mathematics:

> *With additional funding, the mathematicians designed a program to help teachers design projects that could be incorporated into the high school mathematics curriculum, with an initial focus on Algebra II, Geometry, and Trigonometry. Over the next two years, an increasing number of Desert View teachers chose to become involved in the writing, editing, and implementing of projects, extending the project approach into Calculus, Algebra I, and remedial classes. By the fall of 1993, every member of the department used projects in their classes, to varying degrees, and approximately one hundred projects were on file in the high school. The projects ranged in scope from developing construction plans for an arch that spans the Rio Grande River to determining the best day to play a football game by using biorhythm functions.*

Although the mathematicians supplied the information about the project approach, the Desert View teachers had been talking about reforming mathematics teaching and learning at their school and they viewed themselves as partners with the university educators. These perceptions match those expressed by the university professors who served as principal investigators for the funded project. It became apparent to the documenters that no single visionary leader had instigated and nurtured the changes. The mathematics teachers at Desert View High School and the mathematicians from the state university had developed a collaborative partnership in which each participant contributed to the direction this partnership would take. At times, different individuals assumed a significant role, depending on which skills (e.g., organizational, consensus-building, problem solving) a particular situation demanded. At other times, because of group cohesiveness, no single individual appeared to be operating as the leader in bringing about change in classroom practice.

The documenters observed further:

> *As outside documenters, we perceived significant teacher ownership of the program—a perception shared by university mathematicians, who tended to refer to the high school teachers as "the experts in implementation." In contrast, Desert View teachers conveyed the perception that university personnel continued to play the leading role in program design and development. These same teachers, however, commented on the need for the mathematics professors to understand and appreciate more fully the daily reality of public school life—"the nuts and bolts of teaching," noted one district administrator, "everything from how classes are organized and how kids behave, to what the textbooks are like."*

A number of teachers expressed appreciation for the fact that the mathematicians devoted significant time to observations and interactions within the high school classrooms. The university mathematicians' periodic presence in the classrooms is in itself important—even as we consider that their role rarely extended beyond an unobtrusive demonstration of support for each teacher's efforts.

Despite the varied perspectives of high school and university personnel, individuals in both settings noted that, once in the hands of Desert View's teachers, the project approach assumed a pace and pattern of implementation different from that which existed at the university. The university's definition of a project as a challenging "multi-step, multi-day, non-traditional mathematical assignment," for example, assumed a more comprehensive interpretation in the high school. As stated by Desert View's mathematics department co-chair:

> *The process of incorporating projects into mathematics classes does not just mean giving students longer or more difficult problems. It is a change in the way students do mathematics and science. Since students work in groups to analyze a problem, brainstorm answers, break the problem into manageable pieces, and arrive at a solution to which all can agree, the process mimics many job situations. Since projects help students teach themselves and each other, the material learned becomes a part of the student, and this gives the student a stake in his own education. The fact that projects extend and apply the students' knowledge and require them to write clearly and explain their solution means that students remember what they learn and see new connections. The real-life applications of projects awaken many at-risk students to the fact that their education has some bearing on life outside the classroom.*

University personnel whom we interviewed readily discussed the fact that they were mathematicians, not mathematics educators, and admitted to lacking the insights necessary to work with students in the high school setting. One of the teachers who initiated contact with the university made this frank assessment:

> *They told us what they were doing at the university and I didn't like it—and I still don't like the way they do it exactly for my classes because I have younger kids and I have to do it differently.... Out there [at the university], for example, they don't do a story line, which is something most of us do here to make the projects more interesting. I try very hard to write projects that the kids can relate to.*

We observed that because of a common interest in improving the way mathematics is taught and learned, the Desert View mathematics teachers and the university mathematicians formed a partnership that resulted in a shared-leadership model for the reform effort. The teachers at Desert View, as well as several other schools discussed in this chapter, noted that the cohesiveness of their mathematics department was an essential component of their reform efforts. Is it possible to implement reforms at a school where this cohesiveness among teachers does not exist? How might a community atmosphere be developed? What might be the effect of a strong visionary leader on a group of teachers, such as the mathematics faculty at Desert View or East Collins? What might be the problems?

COMMON THREADS IN SUCCESSFUL LEADERSHIP

A number of common threads can be found among the R^3M sites that were successful in bringing about some change in mathematics education in their school or district. Two of these commonalties stand out as essential: first, although some form of leadership is necessary, teachers must feel comfortable with the leadership; second, teachers must have a voice in the project. All the visited sites that had created some reforms in the teaching of mathematics and in the way their students learned mathematics demonstrated the common element of leadership. Although the leadership model varied between a single leader and a group of leaders, leadership was a necessary element of the reform effort; someone had formulated a vision for a different way of teaching and learning mathematics and had been able to get others to share the vision. This is not an easy task. As one administrator of a high school observed:

> *Teachers as a group are not good changers. I would venture to say that most of us teach the way we were taught, and that's one reason change is so slow. Change is frightening. You don't always have a model for that.*

In each school discussed in this chapter, the teachers respected the leaders in their school or district and felt comfortable with the style of leadership. In fact, at the sites we visited a second time, more teachers had become involved with the reform efforts; at some sites, like East Collins and Desert View, all the mathematics teachers were supportive of and working to change mathematics education at their schools.

Clearly, an important factor in reform initiatives is the extent to which the teachers have some voice in the decisions that are made. We saw this at many of the sites we visited, but nowhere was it clearer than at East Collins and Green Hills. At East Collins, teachers shared authority—with representatives from the business community, with school administrators, with one another, and with their students—for planning, implementing, and assessing the effectiveness of their change process. As they worked with others to formulate a common philosophy and to decide how it could be implemented, East Collins teachers grew professionally in many ways. Similarly, as teachers in Green Hills participated in their districtwide effort to realize a new vision for teaching mathematics, they shared authority for that change with a variety of other stakeholders. An important outcome of Green Hills's democratic process of developing consensus and shared vision has been a significant contribution to a wide variety of ongoing supports for teachers' professional development.

Some final questions for reflection and discussion include the following: What other forms of leadership can work besides the two models discussed in this chapter? What do they look like? How does a school support continued reform efforts? Does the leadership model change over time or remain constant? How do leaders influence the direction of reform in their schools or districts?

Chapter 5

WHAT IS THE ROLE OF PARENTS AND COMMUNITY MEMBERS?

Karen Graham and Loren Johnson

It was awful; parents weren't willing to accept it because we were actually changing the way we taught mathematics. They were afraid it was New Math that happened after the sixties; they were afraid it was a fad. They were worried and you can understand those worries because we were taking their children; our investment with their children's mathematics education was being tampered with.

—Teacher
Oldburg Elementary School

I never mind when parents come in. I never mind when other teachers come in. If they came in, they would see what was going on and not just say that she's not using textbooks and the kids aren't doing eighty problems with the same goal. But still it hurts when your colleagues say those things in the community, especially when not one second-grade teacher has ever walked through the door in my third-grade classroom.... In our school, parents are allowed to request teachers for their children, so you know the parents always ask their child's current teacher whose classroom they think the child will do best in.... What do you say to parents that say, "My child's second-grade teacher recommended that I not put my child in your classroom because it's too overstimulating and you do hands-on." And you know it hurts, it just really hurts.

—Teacher
Parker Springs School District

Schools are located within, and supported by, communities. Although the parents of school-aged children might have a more immediate concern about what is taking place in the schools, schools find themselves accountable to the whole community. The whole community has a stake in whether tax dollars are being used effectively. At the same time, the community benefits from a school that has a reputation for offering consistent and quality education. The role of parents and other community members in the improvement of mathematics education continues to be crucial. The issues are complex because the majority of parents and community members attended school when mathematics classrooms looked quite different from those influenced by the vision for school mathematics described in NCTM's (National Council of Teachers of Mathematics) *Curriculum and Evaluation Standards for School Mathematics* (1989) and the companion documents, *Professional Standards for Teaching Mathematics* (1991) and *Assessment Standards for School Mathematics* (1995). Parents may believe that they have been successful as products of the "old" system—so why the change? In addition, the changes in curriculum and approach may make it increasingly difficult for parents to help their children at home. Parents, after ten to sixteen years of personal experience as students in schools, think that they understand what mathematics instruction should be but suddenly feel inadequate to assist their children at home.

Poor communication between the schools and the community can result in misinformation and myths about what change in mathematics education is all about. NCTM, in its series of *Standards* documents, recognized the important role of parents and community in the process of reform and placed responsibility squarely on the schools. In the Support and Development section of the *Professional Teaching Standards,* Standard 2, Responsibilities of Schools and School Systems, states,

> School administrators and school board members should take an active role in supporting teachers of mathematics by accepting responsibility for ... establishing outreach activities with parents, guardians, leaders in business and industry, and others in the community to build support for quality mathematics programs. (NCTM 1991, p. 181)

The authors of the *Professional Teaching Standards* point out that for community members to be supportive, they need to know what the goals of the program are and what types of support are needed by the school and teachers to reach those goals. They describe the principal as a central player and chief advocate in this process. This charge to schools and school systems raises interesting questions about the nature of the process of involving parents and other community members in mathematics education reform:

- How do we foster a proactive, rather than a reactive, approach?
- How do we help parents understand the changes taking place in the mathematics program?
- How do we turn negative reaction into positive effort?
- With what types of activities can parents become involved?
- How do we achieve a healthy balance between parental and community concerns and program goals for improvement?
- What can we learn from the current debates about mathematics education?

In this chapter we explore how some of the R^{3}M sites dealt with these issues. Several sites took proactive steps to avert communication problems; in some situations, those steps involved entire communities in the change process. Stories from other sites tell how individual teachers and administrators addressed the concerns of parents and the community as the reform of mathematics education unfolded.

CHANGING A NEGATIVE REACTION INTO POSITIVE EFFORT

The opening vignette describes a first-grade teacher's response to the question, How did parents react to all these changes? How would you or your school or district have answered? Would you describe your response as proactive or reactive? Which approach do you think would characterize the typical response of the majority of schools in the country? What can be done to turn this negative reaction into a positive force? The Oldburg School District supplies one example of a response to these questions. To alleviate the negative reaction from the community, Oldburg created a threefold plan: establish a Math Advisory Council to validate the program, offer workshops and classes to introduce the parents to the mathematics that their children were learning, and develop stronger parent-school communication about mathematics. The Math Advisory Council included representatives from twelve area corporations. These representatives took the position not of wanting to look at what was actually going on in the district but rather of trying to reach a consensus on what the curriculum should look like. Oldburg's assistant superintendent explained:

> *The whole first year, we just discussed things. We'd give them data, they'd ask questions, we would have these monthly meetings, and we would all be frustrated because it would seem that we hadn't done anything. But the accumulative effect of that dialogue was that they were starting to understand some of our perspectives, we theirs, so finally they came up with eleven or twelve goals. They were identical to ours, except for one. So, it put to rest how far apart these experts' and our original ideas were. These folks validated what we are doing.... I always had faith in our goals.*

This validation of the district's goals by the Math Advisory Council helped alleviate parental concerns about the new mathematics program. A positive force had been established, and the council was able to move on to other issues. The assistant superintendent explained that the "council lasted three years ... we have used that model three more times in our district at the beginning of an issue, not at the end of an issue, and it is now a model we use in anticipation of change."

The mathematics specialist at Oldburg's elementary school commented that in hindsight, the district wished that it had been proactive and had educated the parents before problems became apparent. Instead, they had to be reactive and tried to educate the parents through such programs as the PTA. A first-grade teacher remembered what happened:

> *It wasn't until Oldburg provided some parent workshops at night for the parents and all of a sudden they are getting the "Oh, yeah!" They are making those connections; they're understanding math for the first time, and they're beginning to realize in order for our children to perform in this new technology age, they need more internalization of mathematics.*

Thirty-five parents attended these workshops in which they experienced firsthand the mathematics their children were doing in school. This opportunity gave parents insight into the value of the changes in the mathematics program.

Even with efforts like those at Oldburg to communicate with parents and involve parents and other community members in the process, pockets of negative reaction still needed to be dealt with. A fourth-grade teacher in Oldburg remarked, "I think most of them [parents] can see the benefits, but we still have some that it is an ongoing battle." A high school mathematics teacher who is the spouse of an Oldburg Elementary School teacher commented that parents were still raising issues about the computational proficiency of their children. The mathematics specialist pointed out the importance of finding every opportunity to demonstrate to the parents the strengths of the new approach. This information can be conveyed through workshops as well as in individual conferences with the parents. Choosing the appropriate language and attitude to communicate with parents is crucial. One possibility might be to emphasize what the students can do rather than what they cannot do, as illustrated in the following description by a mathematics specialist:

> *A parent came to me at one point and the child was leaving fourth grade and then she said, "Well, I think she should be able to do multiplication and I want to work with her over the summer," ... and so knowing that we had done some two-digit-by-one-digit multiplication, I said to the child who was standing there, "Well one of the things that we had done was, if you went into the store, a pencil costs sixty-eight cents, and you were going to buy three of them, how much money would you be spending?" And the child went through it in such a way that it was unbelievable [to her mother] how she got her answer, and her mother just looked at her and said, "How did you do that? I could never!"*
>
> *The mother saw there was value in the idea that the thinking was what was important ... and a lot of that has to go on.... We have*

learned to talk about this with parents much more effectively now. I don't know if we were defensive or what all the reasons were, but we would use language that would fire them up instead of calm them down and reassure them. I think we are much better at it.

What began as a reactive approach to concerns about the changes that Oldburg was making in its elementary school mathematics program developed into a positive relationship between the school and the community. What lessons can be learned from their experiences? If the Math Advisory Council had been established for proactive rather than reactive reasons, would it have had the same effect? In addition to validating the mathematics curriculum, what other roles could such an advisory council play? One of the R^{3}M sites developed a more proactive approach. We examine the story of this site in the next section. In the end, you will find it important to reflect on both stories and to discuss them within the context of your own experiences and your own community.

FOSTERING A PROACTIVE APPROACH

The Green Hills School District began its mathematics curriculum effort as the district was moving to site-based management and involving all stakeholders in the process of decision making within the district. Six parent-group focus meetings helped define the process and set the stage for a collaborative effort among the various participants. The result was broad-based participation in each of four working committees during the twenty-nine months prior to implementing the new mathematics curriculum. Figure 5.1 gives a complete calendar of the curriculum development process. The curriculum guide was written with both the NCTM *Standards* document and the state guidelines as models.

The importance of involving both the parents and the teachers in the process became evident when the curriculum philosophy and goals were unanimously approved by all the Green Hills mathematics teachers a year before implementation. Many noted that the year of doing research and involving teachers and parents helped secure this approval.

To assist the curriculum development process, and to ensure the success of the curriculum vision thus far developed, the research committee also created a set of guidelines. These guidelines addressed six categories: (1) assessment; (2) communication; (3) integration; (4) learning processes, instructional techniques, and learning styles; (5) staff development; and (6) technology. These guidelines spoke not only to the teachers and the students but also to administrators, parents, and business partners.

We can hypothesize that the early involvement of parents and other community members in the process resulted in explicit statements in

Figure 5.1

Green Hills School District Calendar of Mathematics Curriculum Development

A. Current curriculum and trends are evaluated.
(29 months [April 1990] prior to September 1992 implementation)

B. Research committee is appointed.
(25 months [August 1990] prior to September 1992 implementation)

C. Review of research is conducted by research committee.
(24–17 months [September 1990–April 1991] prior to September 1992 implementation)

D. Research committee develops guidelines.
(24–17 months [September 1990–April 1991] prior to September 1992 implementation)

E. Research committee develops philosophy and goals.
(17 months [April 1991] prior to September 1992 implementation)

F. Writing committee is appointed.
(16 months [May 1991] prior to September 1992 implementation)

G. Writing committee develops course objectives.
(15 months [June 1991] prior to September 1992 implementation)

H. Curriculum (philosophy, goals, objectives, scope and sequence) are refined.
(12 months [September 1991] prior to September 1992 implementation)

I. Preview instructional materials are ordered.
(12 months [September 1991] prior to September 1992 implementation)

J. Curriculum is validated.
(11 months [October 1991] prior to September 1992 implementation)

K. Preview instructional materials are sent to schools.
(11 to 7 months [October 1991 to February 1992] prior to September 1992 implementation)

L. Curriculum is presented to BOE Curriculum and Instruction Committee for input.
(10 months [November 1991] prior to September 1992 implementation)

M. Curriculum is approved by Education Division.
(10 months [November 1991] prior to September 1992 implementation)

N. Instructional materials are adopted.
(7 months [February 1992] prior to September 1992 implementation)

O. All district teachers write teaching suggestions and instructional materials correlations.
(11 to 6 months [October 1991 to March 1992] prior to September 1992 implementation)

P. Writing Committee develops parent handbook and a course narrative.
(4 months [May 1992] prior to September 1992 implementation)

Q. Staff development is presented.
(4 months to 1 month [May to August 1992] prior to September 1992 implementation)

R. Evaluation component is developed.
(throughout the first year of implementation)

S. Annual curriculum audit is conducted.
(each fall beginning in 1993)

the curriculum guide about their roles. For example, the communication guidelines state that parents will be given in-service opportunities by the district; that communication lines among parents, teachers, and students will be promoted; and that administrators will become advocates for the mathematics program. Similarly, the technology guidelines state that the appropriate use of technology will be modeled; that each student, teacher, and administrator will use technology on an ongoing basis; and that business partnerships will be explored to bring real-world problems into the curriculum.

After the research and writing committees completed their work with the curriculum guide, the other committees began reviewing instructional materials, writing a handbook for parents, and planning in-service opportunities for teachers and parents. Various types of in-service activities were conducted. A workshop, "Math, Tune in the Future," was developed by grades K–6 teachers to introduce parents to the use of manipulatives in the classroom.

Parents were seated in groups. Teachers from each grade level, in turn, used the same type of manipulative to introduce mathematical ideas at that grade level. For example, parents used rods for counting, for developing an understanding of equivalent fractions, and for writing sentences that looked very much like the algebra they remembered. They had opportunities to ask questions and share their insights. Parents were convinced that their voices mattered when at the end of the session, they were asked to fill out a formal evaluation (see fig. 5.2) to help define future workshops. The role of the students as ambassadors for the mathematics program cannot be underestimated. The results from the evaluations were positive; this was attributed to the children, according to one teacher who stated, "Kids are excited, they talk about math with their parents."

Many of the questions parents most frequently asked about the new mathematics program were anticipated and answered in parent handbooks. In addition, teachers joined a television program that addressed

Figure 5.2

Evaluation
Math ... Tune in the Future

February 2, 1993

1. Do you feel you have a better understanding of the math standards developed by the National Council of Teachers of Mathematics?
2. What is your opinion on using manipulatives and/or group learning to teach the math standards?
3. What other topics would you like to see presented at future math workshops designed for parents?

parents' concerns, spoke about the new curriculum, and gave examples of what a parent might expect to see in his or her child's classroom. In a district newspaper, parents were informed about how calculators were being integrated into the curriculum and how they could buy calculators at the district's discounted price.

The process by which the new mathematics curriculum guide was developed at Green Hills was perceived by the teachers, parents, and principal as a model for change. No one seemed to feel left out of the process. In general, administrators, teachers, parents, and students looked forward to increased involvement in mathematics. Green Hills appears to have accomplished its goal.

The Green Hills experience raises some questions about the process of involving the whole community in reform: What were the factors that led to the changes that took place? What do you see as the strengths in the process described here? What are the weaknesses? What problems do you think they might have encountered? What about the Green Hills story makes you skeptical? What questions would you like to ask the people involved? How was the process similar to, or different from, what happened in Oldburg? Would a process similar to the one adopted by Green Hills work in you school or district? Why or why not? What would your school or district need to undertake such a process?

HELPING PARENTS UNDERSTAND THE CHANGES

The previous two sections described districtwide efforts to involve parents and other community members in the reform process. Experience suggests that communication on a smaller scale—between parents and teachers, between parents and students, and between parents and schools—is also important. The NCTM *Standards* documents have caused teachers and school communities to rethink not only what mathematics is taught but how it is taught. Parents, through their own experiences with schooling, may be used to different definitions of what school mathematics is and may have different ideas about what types of homework and in-class experiences are appropriate. Students may also have had similar experiences; as a result, the message that is communicated to parents is that mathematics is *not* being taught. This conclusion may be the extreme, but teachers to whom we talked believed that it could become a reality if careful thought were not given to helping parents understand and appreciate the changes that are taking place in many of the country's mathematics classrooms.

One way of dealing with this problem was to put at the top of mathematics papers going home an identifier indicating that the paper was the student's mathematics paper for the day. On days when mathematics manipulatives were used, the teachers found that students would frequently report to their parents that they did no mathematics that day. A district mathematics specialist in the Oldburg School District reported that teachers averted this miscommunication by reminding students, "This is math class today; this is what you are doing in math."

Sometimes it is helpful to write on the back of homework papers the mathematics involved. For example, a parent might find "Area = 12 square units" written on the back of a student's tessellation design. This notation would then prompt students to explain the mathematics involved to their parents.

Some teachers view the changes in school mathematics as an opportunity to communicate new information to the parents. An elementary school teacher interviewed after an observation in conjunction with an R^3M visit expressed surprise at the level of understanding of her first graders. She was very pleased with the different number of shapes that the students found. She believed that she knew her students better than ever before because they could tell her how they were thinking and learning. "I can report to the parents much easier all the different avenues of mathematics. Before, all I could ever report to them was how they were doing on their calculations, and that was it."

Other teachers considered portfolios vehicles for communicating what mathematics the students were working on as well as their individual growth. One teacher at East Collins had the students save all assignments and activities. Near the end of each six-week grading period, she instructed her students to select some of their papers on the basis of certain criteria. She also selected some that she wanted each student to include. The students completed a self-assessment and had their parents complete an assessment. The teacher also assessed this collection of papers; her assessment counted for 10 percent of each six-week grade.

A second-grade teacher at another site described the portfolios her students were keeping:

> *Each of the students has a three-ring binder divided up by subject areas, and each subject has a table of contents. When I return work to them, they enter the date and the title in their table of contents in the subject area. Parents are always welcome to come in and see their three-ring binders. We have to negotiate where we put the material. There are a lot for subject areas. What is it? Is that reading or math? To me, that's a compliment that everything is that integrated. They put their work in, and then at the end of the nine weeks, they do a lot of self-evaluation. All they have to do is look at their table of contents to see. We staple two masterpieces that they have chosen to their table of contents for that nine weeks in each subject area, and then put those in their portfolios.*

In addition to the portfolios, the students wrote newsletters every week to bring home to their parents to let them know what was going on in the classroom. The newsletter had to be signed and returned on Monday.

Portfolios, parent newsletters, and notes at the top of worksheets are all ways to communicate the types of mathematical experiences that are taking place in schools today. Another means is to involve the parents directly in the activities that are happening in the classroom. At Southland Elementary School, the faculty took advantage of a construction project to expand the school to create a mathematical learning experience for all. In the year before the R^{3}M visit, the school had undergone extensive renovation, doubling its physical space. The faculty at Southland decided to use the construction process to teach a schoolwide unit called "Construction Math." Administrators, parents, and teachers all participated in the teaching of this unit. A teacher described a parent's participation:

> *This parent spent a good bit of time in the bathrooms with the teacher and the kids. We'd send the boys in the boys' room, the girls in the girls' room, and they had to measure the bathrooms and the sinks and how wide the stalls were. Then they had to go back and do their scale drawing, and they actually used clay and modeled the bathroom. We used transparencies to show the plumbing. We used straws to show the plumbing underneath.*

After two days filled with observing these types of activities, the R^{3}M documenters commented that the students appeared to enjoy themselves while they engaged in mathematics activities. A teacher concurred: "I think the attitude has changed. The attitude of the children has changed toward mathematics, and the attitude of teachers has changed. Math is more fun now. You know, it's not just like, 'Oh no, it's time for math.' They're excited about math and it's real to them."

A parent agreed:

> *I know that the program here is designed to teach kids math by being involved with math and letting them understand that math is part of everyday life—not one hour in school, five days a week. This is very important to my husband and I.*

A teacher recounted that she had directed parents who were taking their child out of school for a week to visit Disney World to take advantage of the travel time:

> *When you get on the plane, get her [the student] to talk to the pilot; talk about how far she's going to go up in the air; let her see the controls; let her listen to the pilot as he talks about the elevation, the height at which they are flying, how fast they are flying. When you get to Disney World, let her tell the cashier how much money she should receive back.*

The teacher related how the student wrote a mathematics journal entry for each day she was out of school and that, according to the student, "things just automatically popped up that you can do with math."

Having parents support this view of mathematical learning was important to the staff. Another teacher noted,

> *I think many of our parents realize that we're trying to give their kids a footing in the real world. The math we're teaching really is relevant to the kind of math they need to know to be successful citizens, to be successful consumers, to be successful in their jobs, and to like math.*

These views appear hard to argue with, yet debates exist. All teachers want the best for their students, but they disagree over what "the best" is and how it should be accomplished. Parents want more say over the content of what is taught in schools. Teachers are frustrated because they believe that they have the expertise and should be allowed to make the decisions. One teacher, when discussing related issues, expressed his frustration in this way: "We're not antiparent; I'm afraid this might be coming across as being a little negative—they don't know enough to make the right decisions. That's not exactly what we are saying." Given this tension in many schools over who the "experts" are, how does a school community or an individual teacher achieve a healthy balance between parental and community concerns and program and personal goals for improvement? In what ways can a teacher communicate to parents not only an individual student's growth but also the types of mathematical experiences that occur in a classroom? How can the value of these activities be communicated? What are the major concerns in your community? How would you rate the level of parental concern in your classroom and school district? How have you or others in your district dealt with parental or community concerns? What advice would you give to others?

THE PUBLIC SIDE OF CHANGE

The NCTM *Standards* documents discuss changes in *what* mathematics should be taught in schools as well as changes in *how* it should be taught. The documents have frequently raised the level of discussion around many issues in mathematics education. Some of these issues, such as the role of standardized testing and the amount of emphasis that should be placed on computational skills, are not new to the mathematics education community. For many, these very sensitive and political issues have served to polarize communities. A lack of communication and a lack of first-hand experience with what is actually occurring in the classroom have led to misconceptions and radical mandates.

Many of the visited sites were struggling with these issues. The second vignette that opened this chapter describes one ramification of the struggle. The teacher in this vignette was frustrated that colleagues, who did not share her vision of mathematics teaching, were communicating their distrust to parents by suggesting that they not place their children in her class. The assistant principal recognized and acknowledged the tension with the faculty. However, she believed that the teachers should take a leadership role in any change within the school and that change should not be directed from above. The assistant principal believed that there was a definite need to look at things in different ways, especially in the second grade, where she had heard more than one parent complain about the lessening of a child's interest in mathematics after kindergarten and first grade. She thought that some teachers were holding back and needed to network and share ideas. But she wanted this collegiality to occur naturally because it would be a more powerful way to build commitment.

The teacher in the vignette was one of the district's mathematics specialists. She explained that part of her role was to model a different kind of instruction and a different kind of content. Being a model is a two-edged sword to the extent that what is modeled is public; although the visibility bears the promise of facilitating change, it also has its risks.

Some parents, however, were quite vocal in their support of the changes that were occurring in the mathematics program. The one concern that was expressed by several parents was the lack of continuity from year to year as students changed teachers. The parents of a second grader noted that their son used to "live" mathematics. Every dinner-time conversation related to mathematics. Now, the second grader did not talk about mathematics and, more important to the parents, did not like mathematics in school. Although these parents expressed concern about more consistency from grade to grade and believed that skills were important, they believed that their child's hunger for learning mathematics was a greater concern.

Another R^{3}M site dealt with the issue of parental choice by not offering one. When teachers were asked if they had parents who wanted to move their children out of the program, their response was, "Yes, but they have nowhere to go. You can go to any other class but they are doing the same thing." One of the teachers at East Collins High School told the following story:

> *I would hate to think what would happen if we'd have half of our algebra 1 classes [PSML] and the other half the traditional way. The first week of school, literally, I had a parent call me and say, "My son can't learn that way," and since there was no place for him to go, he stayed and he made Bs and Cs and I have him in algebra 2 this year and he's fine. He's never made As in algebra—or anything else for that matter—but he's doing fine. He's doing at least as well as he ever did before. I haven't heard from that*

parent since the first six weeks last year, when he got that first B. She was very pleased.

This site believed that teachers were the crucial elements, according to one of the school administrators at East Collins:

The teacher is the main piece of it, and that is justified when you put in a brand new program that depends on teachers' being responsive to parents, responsive to criticism, responsive to sudden changes, responsive to long waits. We had to wait sometimes for pieces of equipment or books; you just need teachers who can adapt or change, or it's not going to work.

Another issue that can create controversy and strong debate within school communities is the issue of standardized testing. An example is seen in the story of the Manzanita Middle School, where the superintendent considered that he was trapped in the difficult position of trying to balance the political and parental pressures to explain declining test scores occurring at the same time as major educational reform. He found it imperative that he be guarded in commenting about the testing issue to avoid misinterpretation by parents, teachers, and administators.

As was the situation with computational skill, the value of standardized testing was debated among teachers and administrators as well as among the community at large. In spite of the superintendent's frustration with the misalignment between the goals of the Standards and standardized achievement tests, he was optimistic that an assessment plan could evolve that would meet the needs of children and also yield useful information to teachers and parents.

What would such an assessment "system" look like? What information would you want it to provide? Would you want it to be norm referenced so that you could compare your students with others, or criterion referenced so that you could tell what your students can do? Which do you think parents would prefer? Why? What types of assessments are you currently using? Does your school give standardized tests? What type? How is the information used? How is the information communicated to parents? What would you like to see changed in the way the test is administered? In the way the results are communicated? Does your state require a statewide assessment? If not, do you think one would be a good idea? Who should be involved in the development process?

SUMMARY

Schools and parents share responsibility for raising the nation's children. It is not surprising that problems arise when the goals of these two groups appear to be at odds. Recent calls for educational reform have brought the debate between schools and parents to the fore in many communities. Many parent groups appear uncomfortable with the notion of reform. They are distrustful of educational innovations that appear more like fads than thoroughly tested theories. They want the best for their children. Parents are worried that any change may give their children a less than quality education and deprive them of the best postsecondary and career-related experiences. Parental concerns range from worrying which teacher or textbook will offer the "best" experience for their child to feeling that an emphasis in mathematics on open-ended problem solving gives students too much freedom. Parents in many communities are calling for a stronger voice in decisions about the content of what is taught, how it is taught, and how students are assessed. On the other side are the teachers, who believe that they are the content and education specialists. The teachers are frustrated because they are not afforded the same level of professional respect as are, say, doctors or lawyers. They believe that they have the expertise and experience to know what skills and concepts are essential—and what experiences are crucial if students are to develop an understanding of these issues. Somehow the lines of communication need to be open, and communities need to work to depolarize the debate. For efforts to change mathematics education to succeed, we must work to change schools into places where these debates can occur without risk to teachers' jobs or students' learning. This chapter has presented some examples of how several R^{3}M sites have dealt with this issue. For many of them, it was only the start of a process that needs ongoing attention.

Chapter 6

WHAT ROLES DOES ASSESSMENT PLAY IN MATHEMATICS EDUCATION REFORM?

Diana V. Lambdin

> *Our [standardized] test scores just came back and they're terrible.... Our testing just doesn't match the [*Standards*-based] curriculum.... We end up trying to explain that the achievement tests are not measuring what we want to be teaching, and it puts us in a hell of a bind.... The reality is that I'm deeply concerned about this year's assessment and the translation that people will make to this reform effort that we have under way. It has the potential of derailing what we're doing. It's that caliber of problem.*
>
> —Superintendent
> Manzanita School District

Assessment undoubtedly plays a variety of important roles in mathematics education reform, although whether assessment promotes or inhibits change depends in large part on when, where, and how we look at the issue. We can find examples where new emphases in curriculum and instruction have produced changes in assessment practices, as well as situations where the influence seems to have gone the other way, for example, where troubling test scores have resulted in calls for changes in curriculum and instruction.

Indeed, it is certainly no exaggeration to credit assessment with prompting many of the recent widespread efforts to change curriculum and instruction in mathematics in the United States. Dissatisfied with U.S. students' mathematics performance on national and international tests, society in general—and the education community in particular—have called for change.

> International comparisons of student performance in mathematics show that U.S. students lag far behind their counterparts in other industrialized countries.... The time is ripe for a new approach to curricular reform in the United States—one that establishes appropriate national expectations based

upon broad public support by all of the constituencies concerned. (National Research Council 1989, pp. 8–9)

Clearly, assessment can serve as a catalyst for change.

However, assessment can also be seen as an inhibitor of, or even a deterrent to, instructional reform. In particular, when established assessment schemes are misaligned with reform agendas, students' poor test performances can threaten to derail reform efforts. It is no simple matter to figure out how to document and evaluate students' growth with assessments that are at the same time acceptable to parents and administrators and appropriately aligned with modified instructional methods and curricular changes. Teachers faced with this challenge sometimes become disenchanted with the reform that they perceive as putting them in such a bind. In such situations, we see—in several ways—the potential for assessment to inhibit educational change.

These various perspectives raise important questions and issues about the relationship of assessment to mathematics education reform:

- To what extent are recent changes in mathematics curriculum and instruction related to low standardized-test scores?
- What are the potential benefits and drawbacks of a mandated administration of standardized tests?
- How can district administrators, principals, teachers, and the community at large cope with the threats that traditional tests can make toward derailing reform efforts?
- How can educators at all levels (classroom, district, and state, etc.) assess students in ways that are more appropriately aligned with new emphases in curriculum and instruction? What do we mean by such alignment?
- To what extent should assessment systems focus on gathering data for comparing students (or programs or schools) with one another? To what extent should they focus on gathering data to describe what students know and are able to do? In what ways are these goals at odds with one another? In what ways are they compatible or complementary?
- How can assessment be used as a catalyst for change rather than as an inhibitor?

This chapter discusses two primary ways, identified above, in which assessment has an impact on education reform efforts and illustrates each through stories from the R^3M sites. The first section of the chapter looks at how assessment sometimes acts as an impediment to change. This section focuses, in particular, on problems encountered when mandated assessments (or public perceptions of what students should be accountable for learning) are misaligned with the thrust of educational change. When this happens, assessments may derail reform efforts or at the very least, provoke considerable consternation, confusion, and debate until instruction and assessment are brought into better alignment. The second section of the chapter examines examples of the more positive situation in which assessment is credited with promoting change, not inhibiting it. In these examples we see how changes in assessment at levels beyond the classroom may encourage or even drive reform in instructional practice. The third section identifies four important issues that arise when we examine how assessment and reform issues intersect and discusses how these issues reflect a new and coherent vision of mathematics teaching and learning. Finally, the fourth section challenges readers to explore how this vision can be realized in their own instructional and assessment practices.

ASSESSMENT AS AN IMPEDIMENT TO CHANGE

The opening quotation in this chapter comes from a superintendent faced with a serious threat to his district's curricular and instructional changes in mathematics. The threat came from dismal scores on the state-mandated California Test of Basic Skills (CTBS). The Manzanita School District had made a major commitment to implementating the *Curriculum and Evaluation Standards for School Mathematics* (NCTM [National Council of Teachers of Mathematics] 1989). However, on the day that R^{3}M documenters visited to interview the Manzanita administrators about their efforts to improve mathematics teaching, the superintendent arrived late to the meeting because of a radio interview during which he had been asked to interpret and explain the declining CTBS scores to the public.

Manzanita Faces an Assessment Crisis

In the pages that follow, I examine the Manzanita School District's problem with assessment from the perspectives, interpreted by the R^{3}M documenters, of the central administration, the high school principal, the mathematics department chair, and the teachers.

The Central Administration's Perspective

When questioned about the district's testing program, the superintendent responded in terms of the curriculum and instructional changes in the district and his inability to explain the terrible test scores they had just received. The assistant superintendent proceeded to explain that the district had a team of teachers who had been working for two years in the areas of mathematics, literacy, and work skills to develop and pilot an item bank of alternative assessment items. However, in spite of this local assessment work, the district was still held accountable to a state mandate requiring the administration of standardized tests, specifically the CTBS. The superintendent explained the procedure:

> *The state buys the test and sends it to us. We administer the test and the results are sent back to us.... I'm working with the State Commission on Student Learning, which is charged with legislation passed in March of 1992 to devise, develop, and prepare for implementation in 1996–97 a performance-based assessment system for the elementary schools ... and to prepare for a 97–98 implementation in the secondary schools.... The performance-based assessment is also bringing up a lot of right-wing resistance and antichange resistance and interest-group resistance. They want detailed answers before they make any investment in making a shift and we don't have a lot of that.*

The central office administrators perceived that the district was in a difficult transition period as it struggled with the mixed messages it was receiving: test scores were declining at the same time that the district had just made a major commitment to the implementation of the NCTM *Curriculum Standards.*

> *The message we tried to give our principals is that while we have some questions about why there is a dramatic change in our test scores over the last year, we don't want to transmit the message that "Oh boy! They're really worried about test scores." That isn't the question. The question is, what causes the change?*

The superintendent believed that the posturing over possible causal relationships between curriculum and instruction changes and the decline in test scores was problematic for the district's teachers, principals, and administrators.

> *We're immediately placed in a difficult position, and we end up trying to explain that the achievement tests are not measuring what we want to be teaching, and it puts us in a hell of a bind. And the bind is this: A state legislator yesterday at this meeting said, "We need to keep this statewide standardized achievement test that we are currently doing, even after we move to performance-based assessment because it's what people know and understand." This guy is not a major party senator now, but he was, and he's a former school board member. He's been part of the governor's Blue Ribbon Committee on Education for the past eighteen months.... The reality is that I'm deeply concerned about this year's assessment and the translation that people will make to this reform effort that we have under way. It has the potential of derailing what we're doing. It's that caliber of problem.*

The assistant superintendent was quick to point out that the decline in test scores was a problem not only in mathematics but also in reading and language. Declining scores in areas not specifically related to the implementation of the NCTM Standards were particularly worrying to the superintendent: "On the one hand, I think, 'Is this an indicator that what we're doing isn't the right thing?'" However, the superintendent felt trapped in a difficult position as he tried to react to political and parental pressures to explain why declining test scores were occurring at the same time as major educational reform:

> *Generally, I've been downplaying the significance of the standardized nonreferenced achievement tests for several years.... I mean, I have communicated distance from the process, and five years ago the district was putting incredible pressure on administrators to raise test scores. So we've been through this kind of shift, and last year we began a systematic process of saying that in the spring, schools only have to do selective tests, ones which are judged to be more in line with thinking skills and the* Standards.

The superintendent believed that Dr. Smith, an educational consultant who was helping the district with its reform efforts, had been helpful in communicating to teachers and administrators the problems associated with the use of norm-referenced assessment and multiple-choice examinations and the conflict that arose from using these measures to assess student outcomes in Standards-based mathematics instruction. The assistant superintendent also shared concerns about the interpretation of the test scores, not only in terms of how the community viewed the results but also with respect to the messages that were being sent to teachers. He was aware of concerns that part of the problem with the declining test scores might be attributable to the attitude of the teachers, who felt that they were just playing the testing game:

> *If that's the attitude we're going to take and if in fact it is going to affect our public support for education, then I think there are some things we have to explore so that we don't just say, "Well [the test scores] are not important to us, so we're just going to play the game." They are a requirement, and we need to figure out how to do it with quality and not drill and kill.... I do think that there are some test-taking strategies that we could be using with our students and don't think that's cheating. I think that applied learning is what we're talking about with mathematics.*

Finally, the superintendent summarized the challenge associated with the use of standardized-test scores by explaining how the test scores had become a leverage point for promoting reform and change. It was the superintendent's perspective that he needed to be very careful about what he said about the testing issue for fear that his comments would be misinterpreted by the teachers, parents, and administrators:

> *Part of what we're going to do is to be as aggressive as we can on advocating for what we want and what we believe in. I don't*

want to back away from what we do. I wish we were a little healthier where we could have a more scientific discussion about these standardized tests so we could look at the test results and say, "Well maybe there's something of value in there for us. What is it and what is it saying?" We've been there and they go ballistic when you start having that conversation, which to me demonstrates how assessment is the leverage point. We've got to change.

The Principal's Perspective

The principal at Manzanita Middle School reinforced the concerns about using standardized-test scores to assess student outcomes in mathematics. Like the central office personnel, she worried about misinterpretations of the results by teachers and parents. Although the principal supported the mathematics education reform effort, she was concerned about the decline in computation scores on the CTBS. She thought that the most important thing the district could do to support the reform would be to abolish standardized testing altogether:

I don't think there should be standardized testing because it colors, it sort of taints what we're viewing.... Even within this district, the middle school that's in the higher socioeconomic area always has the higher test results.

The Mathematics Department Chair's Perspective

Mr. Roper, the mathematics department chair at Manzanita Middle School, considered assessment the most troublesome aspect of attempting to implement the *Standards* document: "Assessment is so difficult because we're so used to putting in an objective quantity or an objective grade. That's really not an alternative anymore." Mr. Roper believed that he and his colleagues were facing a problematic period of considerable change. The state had mandated the use of performance assessments by 1996—and these would, presumably, be better aligned with *Standards*-based instruction—but in the interim, teachers were still required to administer the CTBS.

To prepare for the move to performance assessment, Mr. Roper had been attempting to use some "continuous assessment" techniques to evaluate "where the kids are at." However, he expressed concern about his effectiveness in providing meaningful feedback to students and to parents. In fact, on the basis of his own experience with teaching algebra at the middle school level, Mr. Roper believed that the standardized achievement-test scores were accurate predictors of which students would do well in algebra. As he put it:

I've always felt that the standardized tests correlated well with student performance in class, and we use them as one of the criteria for screening kids into algebra.... I find that down to about the 80th percentile, students do well in algebra.

Mr. Roper believed that the teachers at Manzanita were in a time of change and that it was "okay to make mistakes" along the reform path. Although frustrated with the apparent mismatch between the standardized achievement tests and the goals of the *Standards*, Mr. Roper was positive about the eventual development of an assessment "system" that would give teachers, children, and parents more useful information.

Teachers' Perspectives

The teachers at Manzanita Middle School were unified in their belief that using standardized-test scores to evaluate their students' achievement did not make sense. In the teachers' opinions, the tests lacked validity—that is, they did not test what the teachers were teaching as a result of implementing the *Standards* documents. Consistent with their perception of being "pioneers," the teachers played down the utility of the test scores by viewing the tests as outdated. As Mr. Bell, a teacher at Manzanita Middle School, commented:

> *I think that basically the math teachers don't pay much attention to the test results that we got. Someone mentioned that we were in the 31st percentile or something like that.... But my math colleagues know why that is. We don't teach computation. The kids were not allowed to use calculators on the test. It seems like really antiquated ideas that are being tested and we don't feel bad about that at all.*

Mr. Bell believed that the only faculty member who had expressed concern about the test scores was the school's eighth-grade counselor, who had come into the classroom to check whether the children had been "messing around" during the test. Mr. Bell had conveyed to the counselor his perception that the children had taken the test seriously and worked well during its administration. Even though the counselor expressed concern about the test scores, Mr. Bell believed that the message not to be concerned about test scores had been transmitted by the school and district administration, who had "said nothing" about the scores.

Another teacher at the middle school expressed similar views on the use of standardized tests. Ms. Wheeler said that the tests had no effect on her teaching; they were just something that the children had to do: "I just tell the children to take the test, but I don't teach to the test. I don't even look at the tests ... and I've never heard anything from the principal about the tests." According to Ms. Wheeler, the administration of the tests was simply an example of the state's testing policy not matching the school curriculum. It was her perspective that the test results lacked any practical application to her teaching. She was unclear about what the administration did with the scores:

> *I really don't know what purpose these tests serve.... Are they just for comparison? No one here has said, "Gee, your kids have decreased in computation skills, so is your new math program really working?" I don't hear much about the test results. They're not shared with us in the sense that ... this is what the test showed so you should be doing more of this.*

Standardized-Testing Issues at Other R³M Sites

Manzanita was certainly not the only R³M site where standardized testing was considered a serious issue. At Bedford Middle School (BMS), the content of the State Mandated Test (SMT) had been in flux for several years, which left the teachers very confused about what would be tested and what should be emphasized in classrooms for students to do well on the test. Those teachers who were active participants in the Mathematics Teaching Project (MTP) (see p. 26)—the district's effort to implement the NCTM's Standards—claimed that they endorsed the *Curriculum Standards* and a problem-solving orientation to teaching. But they were reluctant to fully implement Standards-based ideas in their classrooms for fear that student scores on the SMT would suffer. Unlike the Manzanita teachers, who professed not to worry about their students' standardized-test scores, the Bedford teachers could not afford to ignore the SMT. BMS had had such dismal SMT scores in the previous year that the school was being threatened with closure by the state unless the scores improved. To complicate the problem, the previous year's SMT had originally been advertised as allowing calculators for the first time. As a result, some Bedford teachers permitted their students to use calculators freely during the year and structured their instruction accordingly. Then, at the last moment, the state decided not to allow students to use calculators on the SMT, a decision that these teachers believed unfairly handicapped their students. As a result, the Bedford teachers decided not to allow calculator use during the next year. To add to the confusion, no one seemed to know what would be emphasized on the next SMT, although the test was purportedly moving toward more alignment with the *Standards* document.

Similarly, Southland Elementary teachers decided—because of the pressures of standardized tests—not to move too far from their traditional emphasis on building computational skills through practice and memorization. Teachers at Southland noted that they had "studied" the *Standards* document, and they claimed that they valued decreased attention to rote skill development. However, they still were obliged to administer a basic-skills achievement test on a schoolwide basis. The teachers agreed that assessment was a large concern that was just beginning to be addressed at their school. Until the issue was resolved, the teachers thought that they could not ignore "basic-skills practice."

ASSESSMENT AS A CATALYST FOR CHANGE

In contrast to places where assessment seems to have worked against efforts to implement Standards-based instruction are situations where assessment can be credited with promoting, rather than inhibiting, change. Examples of such sites in the R^3M project are Oldburg, Green Hills, and Pinewood. In all three, changes in assessment at the state or county level were seen as supporting or even driving instructional practice that encouraged conceptual development.

How Assessment Acted as a Catalyst for Change in Oldburg

Personnel in the Oldburg School District believed that their state mastery test actually had acted as a catalyst for change. According to the Oldburg mathematics specialist, changes in testing and in instruction had been under way in Oldburg for years—well before publication of the NCTM *Standards* document (NCTM 1989). In the early 1980s, the district was ready to revise its curriculum. The commissioner of the state department of education warned all districts that mastery tests were coming and that they should be aligning their curriculum with such tests. Oldburg, through a state department contact, had access to drafts of the tests—tests that the Oldburg mathematics specialist was convinced "measure more than straight computation." In addition, Oldburg had drafts of the *Curriculum and Evaluation Standards* (NCTM 1989) before the book was published. As a result, the Oldburg district worked to align its curriculum with the *Standards* document and with the new state mastery tests.

The mathematics specialist told us that Oldburg's district Math Advisory Council was a big help in dealing with standardized tests; the council was kept up to date on all the testing data. "In this state we have a mastery test and one of the big areas is mathematics. The fear from some of the people was there was going to be a precipitous drop in the mastery test scores." In fact, the first group of students to go through the program did have a drop in computation scores on the sixth-grade test.

> *There was a great deal of concern, but there were some areas that indicated that higher-level thinking was really improving.... This year the scores really rebounded very nicely, even in areas where there was some concern about weakness.... All of that comes into play—you can't find evidence for supporting your position without doing justice to the statistics of it all. But this [sharing district test information] is something I would encourage a district to do, in terms of setting up such an advisory group.*

Perhaps one important difference between Oldburg and the districts with more serious concerns about mandated testing was that the Oldburg teachers saw their mandated state mastery tests as different from traditional norm-referenced tests, according to the mathematics specialist. The following comments illustrate the positions held by teachers in Oldburg.

> *I think they don't see the state mastery test as being out of line because most of the objectives on the state mastery test were things that were downplaying computation.... And objectives like regrouping for subtraction. Teachers couldn't believe that at fourth grade they weren't expecting kids to have mastered that. So it was almost the other way—like teachers would say, "They can do this. Why isn't that on the test?" So instead of feeling like you had a lot of things that you had to cram into, it was something they could feel a bit more relaxed about, present it all the way along the line.*
>
> *The other thing nice about it is that [previously] many of the times that the norm-referenced test was given, it was given at such a time in the year that the teachers almost looked at it as an evaluation of them, where[as] our state mastery test is [now] given in September of fourth grade.... In other words, they get back here in September, the first of September, and the mastery test starts like September 18th. Well, teachers know it is not [judging] them, so it is a totally different aspect that they go into. Much nicer.*
>
> *Yes, they really do [think the state tests are nontraditional]. And I think one of the reasons they see that is that our assessment with the third-grade criterion-referenced test (that the third-grade teachers really did the development of) gave them again the empowering spirit. They felt challenged that they could develop their own criterion-referenced test, and it was very much a lead-in because it gave them a good indication on where the kids were going to be strong with the state mastery test in the fall and where the kids were going to have problems with the state mastery test in the fall. So they really see the state mastery test as something that gives them information.*

A handbook for teachers distributed by the state's department of education seems to confirm the teachers' perceptions that their state test was more appropriately aligned with their Standards-based curriculum than typical standardized tests were. The handbook states that although the objectives for the test were chosen as "significant benchmarks of growth," limitations on the test format influenced the final list. The major focus of the test is to assess skills at both the symbolic and the abstract levels. The handbook encourages teachers to continue their own assessment of students through classroom observation, the use of manipulatives, and problem solving.

How Assessment Supported Change in Green Hills

State-level assessment also seems to have supported the teachers of Green Hills High School in their efforts to adopt Standards-based instruction. The state had an open-ended assessment test to go along with its Standards-like framework that was intended to drive local curriculum changes. Sophomores at Green Hills were given the Test of Academic Proficiency, which was the only national gauge of how Green Hills students compared with those in other schools and other states. Teachers were just beginning to experiment with alternative assessments in their classrooms. Rubrics were being used by some teachers in the assessment of mathematical tasks, and one teacher talked about having students do an oral test to explain how they used the calculator on problems. Teachers also sought program assessment through interviews with students.

Teacher perceptions of Green Hill's students were positive, although the teachers did view the students as overly concerned about grades, which made alternative assessments somewhat more difficult to implement. Nevertheless, teachers indicated that students were eager and involved and seemed to be enjoying their mathematics. Students were honest in their appraisal of different mathematics activities, which gave teachers valuable feedback to help in adjusting their instructional program.

Mixed Assessment Influence in Pinewood Reform

In large part, reform in Pinewood was initially influenced by concerns about achievement as measured by traditional test scores. These assessment concerns continued to be an important factor in the reform effort there. An early change, now nearly complete, moved the required state test of basic skills—a "fifth-grade" test in terms of content—from the high school to the middle grades. Some time was still being spent at Pinewood High School on this test, but students were now expected to finish that state requirement before arrival. The district also used a locally generated criterion-referenced test for its algebra and geometry courses. Ensuring that those tests were aligned with the changes in curriculum in those courses had been a formidable task for the district leaders, but they hoped that the alignment of their testing program with the reform effort would make assessment a lever that would support, rather than inhibit, change.

A very top-down structure was present in the Pinewood district and at Pinewood High School, in particular. Teachers believed that they must conform or receive low evaluations. However, mathematics teachers were beginning to admit that the reform in mathematics at Pinewood High was starting to work. The county had established a locally developed criterion-referenced test (CRT) to drive instruction to accomplish a mathematics curriculum that supported the supervisor's vision. Teachers were somewhat concerned that the new approaches and content coverage would not allow students to master basic skills, which they believed were still very important and which were included on a statewide functional mathematics test (for which students were coached). Equity 2000 and Pacesetter programs were getting under way, and assessment supplied by those programs was also used. Performance on the CRT had been noted by teachers as acceptable, but students were scoring poorly on the Pacesetter tests. Pinewood's grading policy, which had been very restrictive, was starting to open up as teachers began to employ such alternative assessments as journals and notebooks.

ISSUES RELATING ASSESSMENT AND REFORM

As the stories in the preceding sections show, whether we encounter assessment as an impediment to reform or as a facilitator of reform, new visions of teaching and learning require new ways of thinking about assessment. Mathematics education seems to have experienced a "complete paradigm shift that involves new decision makers, new decision-making issues, new sources of assessment information and new understandings about the nature of mathematics, mathematics instruction, and mathematics learning and problem solving" (Lesh, Lamon, Lester, and Behr 1992, p. 380). Recognizing the source of the impetus for recent changes may help us better understand the complicated interactions of assessment and reform. Lesh and his colleagues claim that the paradigm shift in mathematics education results from four major issues: (*a*) changing assumptions about the nature of mathematics, how it is learned, and what it means to teach it; (*b*) adapting to technology and innovation; (*c*) clarifying the purposes of assessment; and (*d*) establishing standards for judging the quality of assessments.

Changing Assumptions about Mathematics Learning and Teaching

Mathematics educators are moving from viewing mathematics as a fixed and unchanging collection of facts and skills to emphasizing the importance in mathematics learning of conjecturing, communicating, problem solving, and logical reasoning. These changes are related to a trend away from viewing learning as human information processing and toward seeing learning as a complex process of model building. In a parallel sense, mathematics teaching is moving away from lecturing, explaning, and practicing of decontextualized procedures toward helping students construct their own knowledge through the investigation of realistic mathematical problems. These three essential shifts in thinking have also prompted many recent changes in mathematics assessment (cf. Lesh, Lamon, Lester, and Behr 1992, pp. 381–83). But—as we have seen in the foregoing R^{3}M stories—problems arise when changes in mathematics teaching proceed at a different pace than do changes in assessment. The result can often be significant mismatches in philosophy and in practice.

Adapting to Technology and Innovation

New technologies have resulted in significant changes in the real-world problem-solving situations for which mathematics is useful and in the types of knowledge and abilities that are important today. As a result, technology is exerting a major influence on assessment. For example, the availability of handheld calculators and notebook computers with graphing and symbol-manipulation capabilities enables students to think differently, not just faster. Significant problems arise when the use of certain technologies is assumed in instruction but discouraged (or even banned) during assessments (or vice versa).

Clarifying the Purposes of Assessment

Related to the changing views of mathematics, learning, and teaching are the changing views about the purposes of assessment itself. Some believe that "the aim of educational assessment is to produce information to assist in educational decision making, where the decision makers include administrators, policy makers, the public, parents, teachers, and students themselves" (Lesh and Lamon 1992, p. 4). However, traditionally—at least for many teachers—debates about student assessment have been restricted to discussions of grading practices. NCTM's *Assessment Standards for School Mathematics* takes a more inclusive view, identifying "evaluating students' achievement" as just one of four purposes of assessment (NCTM 1995). (Other purposes are monitoring students' progress, making instructional decisions, and evaluating programs.)

Proponents of reformed assessment use qualitative as well as quantitative data, and they focus more on describing student progress than on categorizing individuals or predicting future success. Thus there is a move away from a reliance on multiple-choice tests and toward the increased use of performance assessments, open-ended questions, group projects, portfolios, journal writing, oral reports, and observations. Of course—as we see through the R^{3}M stories—reform efforts can produce a mismatch in content, form, and purpose between classroom assessments and more-traditional assessments (for example, standardized tests). The conflict is not easily resolved because new forms of assessment often are not appropriate for comparing students, schools, or districts—a significant limitation for those who see accountability as assessment's primary goal.

Establishing Standards for Judging the Quality of Assessments

Along with new assessments comes the need for explicit criteria by which to judge their appropriateness and effectiveness. NCTM's *Assessment Standards for School Mathematics* (1995) raises numerous important assessment issues and offers six standards for judging assessments: mathematics, learning, equity, openness, inferences, and coherence. Chief among these issues are deciding what constitutes appropriate assessment for particular situations and determining how educators can learn to use the most appropriate forms of assessments. There are no simple answers to these questions because state departments of education, school districts, teachers, parents, and the public at large all want evidence that students are progressing appropriately, but these stakeholders are often at odds about what they value. Yet even if we take as given certain goals for mathematics education—for example, those in the NCTM's *Standards* documents—deciding how to assess those goals is no simple matter.

As a result, we found at R³M sites many teachers desperate for help with assessment. A teacher from Bedford Middle School, when asked what assistance she needed with her reform efforts, confided, "Assessment—I'd like that ... with these journals, how do I really evaluate them and then transfer all that into a number grade that mom and dad can accept? ... I'd like a lot more with that." Similarly, teachers at East Collins High School realized that changing from using paper-and-pencil tests and grading students' individual work to using other forms of assessment would be a difficult and slow process. One East Collins teacher noted,

> *Assessment is still something I'm working on. I've been worried about that all year. I kind of feel old, set in my ways sometimes. I've been doing a lot of reading lately and thinking about assessment, but I really haven't done too much about it yet.*

THINKING IT OVER—IT'S YOUR TURN

Mathematics education today is in the midst of a major upheaval. As educators at all levels adjust to new visions of learning and instruction, much that we have done or believed for years is being challenged. Similarly, this era is pivotal for everyone involved with designing, implementing, and interpreting innovations in assessment. In instruction and assessment—two complementary educational arenas—teachers, administrators, parents, students, and entire communities are adjusting to change.

Now that you have read some stories about how teachers and administrators at the R³M sites have struggled with assessment-related issues, turn back to the list of questions at the beginning of this chapter. Read through them, asking yourself how they relate to you and your school. What difficulties have you experienced that are related to a mismatch between new instructional methods and traditional (perhaps mandated) assessments? Talk with your colleagues about strategies for keeping this problem from derailing your reform efforts. Have parents (or politicians, or perhaps even administrators) in your community demanded test scores that allow students, classes, schools, or districts to be ranked and compared? Brainstorm how you and your colleagues, perhaps working in collaboration with a community-based task force, can undertake more personalized and descriptive assessment schemes to complement and enlarge the limited vision that norm-referenced tests provide. Individuals, working together in a common effort, can remove the barriers to change that assessment sometimes erects. Indeed, assessment can even become a catalyst for change.

Chapter 7

HOW CAN WE STRIVE FOR EQUITY AND EXCELLENCE?

Loren Johnson

When we said algebra for everyone, we said it, but it was interpreted as "everyone except"—we had this long list of what we were going to exclude.... My understanding is that we really do mean everyone.

—County mathematics supervisor
Pinewood School District

"Algebra for all," and now geometry as well, is a worthy rallying cry for this district. Combating elitism is part of the problem, but the more general issue is changing the beliefs of teachers that doing mathematics is largely a matter of ability rather than effort and opportunity.

—R^3M documenters
Bedford Middle School

Raising standards will not be easy, and some of our children will not be able to meet them at first. The point is not to put our children down but to lift them up. Good tests will show us who needs help, what changes in teaching to make, and which schools need help.

—President Clinton's 1997 State
of the Union Address

A dichotomy exists in educational circles between equal educational opportunities for all students and issues of educational performance. Multiculturists are convinced that both equity of educational opportunity and quality learning can occur in our schools. The *Curriculum and Evaluation Standards for School Mathematics* (NCTM 1989, p. 4) shares this view:

> The social injustices of past schooling practices can no longer be tolerated. Current statistics indicate that those who study advanced mathematics are

most often white males. Women and most minorities study less mathematics and are seriously underrepresented in careers using science and technology. Creating a just society in which women and various ethnic groups enjoy equal opportunities and equitable treatment is no longer an issue. Mathematics has become a critical filter for employment and full participation in our society. We cannot afford to have the majority of our population mathematically illiterate: Equity has become an economic necessity.

Another view holds that the quality of learning suffers when schools seek to provide educational opportunities for diverse student populations. Proponents of this latter position contend that when schools promote programs for a diverse student population, somehow a deficit influence falls over the entire educational system (Ladson-Billings 1992). This position, held by a very vocal and influential minority, is having an impact on educational policy throughout this country. The why-can't-they-learn-as-we-did mentality of this group has been particularly successful in California. Affirmative action has been eliminated in that state, and bilingual education and reform efforts in mathematics are currently under attack.

HOW DO WE ACHIEVE EQUITY AND EXCELLENCE?

Just what do we mean by equity in the educational setting? Is establishing high standards for all students a way of putting equity into education? Achieving equity by raising the educational expectations for all our students seems like a reasonable goal and is promoted by the *Standards* document (NCTM 1989). Ladson-Billings (1994) suggests that providing for equity in our schools is a very complex process and is not something that is easily accomplished. She identifies five areas on which we must focus if we are to address the challenges of offering equitable learning opportunities to a multicultural population: "teachers' beliefs about students, curriculum content and materials, instructional approaches, educational settings, and teacher education" (p. 22). Raising standards is important, but raising standards alone does not present a comprehensive program for dealing with the inequities that exist in our schools.

Albert Shanker, former president of the American Federation of Teachers, often voiced his concern over the inequities that exist in education and believed that it was important to raise standards to achieve equity in education. Shortly before his death, he had posted an Internet message in response to President Clinton's 1997 State of the Union address. His concern was that standards of education in inner-city areas may be vastly different from those found in the suburbs and in rural areas of the country. He was concerned for children in urban communities who had inadequate materials and facilities and for whom achievement levels were extremely low. This message said in part,

As President Clinton says, setting standards will help us help these children. It is the lack of common, high standards for all our students that is cruel and unfair and that perpetuates the inequities in our education system.

The truth of this proposition comes alive in "School System Shock," a terrific essay by an African-American student, which appeared in the school-reform newspaper *Rethinking Schools* (Winter 1996/1997). Melony Swasey, the author of the article, says she had always been confident that she was doing well in her inner-city school. She was in honors classes all through high school and had a good grade-point average—in fact, she was one of the top ten in her class—and she had no doubt that she was headed for success in a good college. But when she transferred for a year to a suburban high school, she found herself in a totally different league—one she hadn't even known existed:

> The classes I have as honors at Kennedy [her old school] are considered only regular college prep classes in other schools. I realized that those of us who are top-notch at Kennedy might be cut down to an average level in more competitive high schools—schools that have higher standards of education.

Weissglass (1996a) believes that achieving equity will require a deeper, interpersonal transformation of those who make up the community called school:

> My goal is to transform schools into communities that nurture (intellectually and emotionally) all the people involved with them. But we cannot transform schools unless we transform ourselves. Personal transformation requires that we develop mutual support systems that can engage us in dealing profoundly and meaningfully with issues that are deeply emotional: how we were treated as learners, how our intelligence was interfered with, how we experienced prejudice and discrimination and saw it affecting others. (P. 4)

Weissglass cautions that bringing about educational change is not a smooth journey and that some of the biggest difficulties we encounter will be those we bring inside us. Even those of us who believe that we have no prejudice may be guilty of acts of prejudice. I, for one, find it difficult not to cheer silently for the African American or Chicano student who does well on one of my mathematics tests.

Others share Weissglass's concerns that we may be emphasizing the wrong things as we go about the task of providing a quality education for all our youth. Noddings (1995) challenges us to think more deeply about our rush for seeking "uniformity in standards" and to seek instead a curriculum centered on themes of care that, she claims, "can be as richly intellectual as we and our students want to make it" (p. 367). Noddings warns,

> [A]nyone who supposes that the current drive for uniformity in standards, curriculum, and assessment represents an intellectual agenda needs to reflect on the matter. Indeed, many thoughtful educators insist that such moves are truly anti-intellectual, discouraging critical thinking, creativity, and novelty. (P. 367)

Noddings (1995) and Weissglass (1996a) both note a need for radical transformations to occur in the way we deal with people. According to Noddings,

> [This transformation] requires organizational and structural changes to support the changes in curriculum and instruction. It requires a move away from the ideology of control, from the mistaken notion that iron handed accountability will ensure the outcomes we identify as desirable. It won't just happen. We should have learned by now that both children and adults can accomplish wonderful things in an atmosphere of love and trust and that they will (if they are healthy) resist—sometimes to their own detriment—in environments of coercion. (P. 368)

By removing "program hierarchies," Noddings believes, we will be in a position to begin offering excellent programs for all children, programs that are rich, desirable, and rigorous for both college-bound and non-college-bound students.

Weissglass (1996a) indicates that to make equity our central focus, we need to build an infrastructure that does not currently exist. Weissglass and his colleagues are contributing to the construction of such an infrastructure through projects that focus on three interrelated components: mathematics, equity, and leadership. In the various workshops presented by these projects, educators develop an understanding of the relationship that exists between their respective cultures and mathematics. Typically, participants in these workshops come from a wide range of racial and cultural backgrounds.

Participants in these workshops discover that every culture has fundamental mathematics activities. They also learn that "any serious attempt to achieve equity in mathematics education must be based on an understanding of how individual prejudices, unaware biases, and systematic society discrimination work" (Weissglass 1996a, p. 6). Weissglass has developed discussion topics that are used in the workshops to build an awareness of common practices in schools and classrooms that perpetuate inequity. Creating this awareness then leads to alternative strategies that when implemented might lessen or eliminate such inequities.

Weissglass (1996a) believes that teacher isolation prevents schools from making progress in dealing with situations of inequity. He says,

> A major obstacle to making progress on equity is the educational culture that keeps teachers isolated from each other and inhibits meaningful reflection or discussion on learning and teaching in general and on how racism, sexism, and classism affect teaching and learning in particular. Teachers need time to talk to each other about how their backgrounds and experiences with prejudice affect them as teachers. (P. 6)

Leadership is the main delivery system through which equitable school communities can be established, and the leadership project directed by

Weissglass is intended to develop leaders who are capable of overcoming resistance to change in the schools. He summarizes the roles and responsibilities of such leaders:

> Leaders will need to understand mathematics, the learning process, and how bias and prejudice work in society and in schools. Effective leaders must be able to develop alliances with different people of different gender, socioeconomic class, and ethnic backgrounds than they are. They will need to be able to provide emotional support to colleagues while helping them rethink some of their assumptions about schools, learning and equity. Finally they must have the skills to raise controversial issues in building unity. (P. 11)

Weissglass is quick to point out that these leadership workshops do not give educators a "package" for equity work. "The process is rather one of increasing knowledge, developing alliances and empowering oneself" (p. 11).

Closely connected to inequities in mathematics education is the condition of poverty that prevails with many underrepresented groups. Access to high-quality instruction is often beyond the grasp of such groups (Haberman 1991; Knapp, Shields, and Turnbull 1995). Haberman (1991) distinguishes between the "pedagogy of poverty" (the teaching that takes place in poor, urban areas) and "good teaching" (the teaching that is found in suburban communities). Teaching methods in poor, urban areas are characterized by strict, teacher-directed, highly authoritative practices in which teachers are afraid of losing control.

Knapp, Shields, and Turnbull (1995) conducted a study of 140 classrooms that analyzed instructional practices intended to enrich learning in high-poverty classrooms. Although teachers in these classes did not always follow fully the current standards (e.g., mathematics) being promulgated by professional groups, they did exhibit many of the qualities endorsed in those standards. These practices generally deviated from traditional teaching practice that emphasized "curricula that proceed in a linear fashion from the 'basics' to 'advanced' skills (though seldom reaching the latter), instruction that is tightly controlled by the teacher, and ability grouping that often hardens into permanent tracks at an early age" (p. 771). It is interesting to note that these researchers found that although teachers in the observed classrooms followed "meaning-oriented" practices, they did not approach their teaching with some packaged process or formula.

MATHEMATICS FOR ALL: A LOOK AT SEVERAL SCHOOLS

The evaluation portion of the *Curriculum and Evaluation Standards for School Mathematics* (NCTM 1989) gives four indicators of a mathematics program's consistency with the Standards: student outcomes, program expectations and support, equity for all students, and curriculum review and change. Although the R^{3}M study was not intended to evaluate a school's alignment with the *Standards* document, documenters were asked to collect data on local perceptions of such an alignment. At the conclusion of a site visit, documenters were expected to address the following areas:

1. The "mathematical vision" held by the people in the site
2. The "pedagogical vision" held, relative to mathematics, by the people in the site
3. How contextual features influence, both positively and negatively, the teachers' efforts to change their mathematics practice
4. The way that the mathematical and pedagogical practices in this school affect students
5. The evolution of the mathematics program in this school

Documenters' feedback on item 4 appeared to shed the most light on issues of equity alignment. Documenters found that through classroom observations and interviews of site participants, they could gain valuable data on how a particular site handled equity issues. The following highlights from the R^{3}M data indicate the level at which several sites dealt with equity issues in their schools.

Higher Expectations at Pinewood High School

> *Equity is a central theme for change in this large urban school district. Since the district is about two-thirds minority students, and the school predominantly African American, equity concerns are an important part of the agenda of the administrators. When we look at our total nonwhite population, it is close to 75 percent, so we really are dealing with some real issues on teaching and learning for all students. Historically, our nonwhite students lag behind white students in achievement, and we firmly believe that it is not an issue of not being able to. More than anything, it is an issue—there are some differences in the way we treat students. I'm speaking from the perspective of mathematics because that is what I know most about. We have historically had an elitist group in mathematics teaching. Who can learn it and who can't and who are the ones who get anointed for academic mathematics and those who are relegated to the hinterlands?*
>
> —County mathematics supervisor
> Pinewood High School

Changing teachers' beliefs was the mission of the county mathematics supervisor. She expressed her concern that one reason for the poor performance of students on standardized tests was the teachers' low expectations.

Raising teachers' expectations of their students' ability to do mathematics was one of her major goals. She explained:

> *If you are identified as talented and gifted, you received a different kind of instruction than everybody else. You got to do exciting things, you experimented, you went on those field trips where you could connect what went on in the classroom with the real world. But if you were everybody else, you didn't have those experiences. So a lot of it has to do with the fact that we've known for a long time what to do and how to do it. We just selected who we were going to do it to.... And what we're trying to do is treat everybody as though they are talented and gifted. Let's treat everybody like they are college bound.*

Almost two years before the R^3M documenters visited Pinewood High School, the school district had decided to participate in the Equity 2000 and the Pacesetter programs (programs designed to encourage participation by underrepresented minorities). During the summer before implementing their reform in first-year algebra, teachers attended a two-week staff development program to prepare for their participation in the Equity 2000 program. Many teachers were cautious about jumping into a program that had been mandated by district administrators; others approached the two-week workshop with varying degrees of optimism.

A particularly enthusiastic high school teacher expressed his impression of the workshop: "That institute we had last summer, it just opened up a whole new world for me. And now when I want to go to these workshops, I realize I can pick up more and help my kids more." This same teacher readily admits that many of his colleagues did not feel as he did about the workshop and about other in-service opportunities provided by their participation in the program: "I think some of that is that people are afraid to try something new. The results really haven't been that good so far. We're always trying new programs in [our] county."

The teachers in Pinewood High School were just beginning to reform their mathematics content and teaching practice at the time of the R^3M visit. Although some teachers showed resistance to the contemplated changes, teachers agreed that reform efforts should continue the following year. These incremental changes in teachers' beliefs about their own teaching practice were seen as monumental steps by some of the teachers in the school. The county mathematics supervisor shared some of her concerns during this time:

> *So my role is not just curriculum and instruction, even though I have responsibility for directing curriculum and monitoring instruction. A lot of what I have to do is deal with affective issues and find ways to say and to make observations about what we're doing and what we're not doing that are sensitive in nature to people who do*

not like to be challenged under their belief systems. That is probably one of the more difficult things that we have to do.

Mathematics Projects at Desert View

Desert View High School is located in the southwestern part of the United States and has a student population that is predominantly Hispanic. When R^3M documenters visited the school, teachers and students had had three years of experience in a mathematics program that required group projects at least twice a year in each mathematics class. The projects developed as a response to a need for more relevant mathematical experiences for students. The group projects, a collaborative effort with the local university's mathematics professors, were seen as giving students meaningful problem-solving opportunities.

The projects had pedagogical implications for the mathematics teachers. Teachers began to recognize their potential for attracting more and more students into the college-preparatory classes. Initially introduced in these higher-level courses, the projects began to filter to most of the mathematics courses offered at Desert View. Teaching practice in most classes began to shift to a more facilitative format—at least during the days when students were allowed class time to work on the group projects. Several teachers, encouraged by the receptivity of students to the projects, began to move away from a lecture-dominated teaching mode. Teachers were enjoying their students more, and students were finding that mathematics could be something besides an obstacle to be overcome.

Reform efforts at Desert View High School resulted in greater numbers of underrepresented students enrolling in mathematics classes that had more challenging mathematics content. According to an assistant principal a "pull-up" of mathematics had resulted:

> *The trend to fewer remedial classes during the past four years from thirty-three to twelve is a positive one.... I think part of this story is that by trying something new and attempting some things that were kind of avant-garde, and getting support and doing it, we were able to, I think, get more kids involved in the math mainstream than before and that it was done without kicking and screaming and crying.*

Interestingly, greater numbers of students were taking mathematics in spite of a declining school population—another strong indicator that mathematics content for students was changing.

Changing Attitudes at East Collins High

East Collins High School is situated in a growing suburban area with a diverse population. Roughly 30 percent of the students are African

American. Teachers and administrators at the school are proud of the accomplishments of their mathematics program. During an interview with the district's assistant superintendent, one documenter asked, "How do you know what the teachers are doing to improve the learning opportunities?" He responded,

> *Well, that's a good question. Let me think about how I can answer that and give you a valid answer. In this particular case, and in a lot of situations, you look at the number of students who are doing well in mathematics, and we are around 57 percent. That's not a significant enough percentage to be satisfied with, and we want to improve that, so we look at the number of students that go on from algebra 1 to algebra 2 and on to higher mathematics. Then you say maybe it's the way we teach them and motivate them; let's do some things to motivate. I think what we're trying to do is to develop a program to get students more involved and more problem solving, and I think that's mathematics. As a result of that, we do have significant increases in the numbers of students that go on to algebra 2.*

One way to reach the diverse student population was to develop a program that would overcome the mathematics phobias that many students had. This goal became a primary focus of the mathematics program at East Collins, a program that received funding from local industry. As the associate principal pointed out, "Students are getting messages from everywhere that say you don't need [mathematics], so it's all right if you can't do [mathematics]." She explained that the program at East Collins High School admits and accepts that students have mathematics phobias. "It became the goal of teachers to encourage students to stay in the math track, and find ways that help students apply math and not just learn it by rote." As further testimony to the effectiveness of the program in reaching a broad range of students, the associate principal added,

> *We went from 57 percent of our algebra 1 students going into algebra 2 the year before we implemented to 96 percent going into algebra 2 the first year after [the program] was implemented. We felt that was an amazing statistic. But what couldn't be measured was the changes in students' and teachers' attitudes. A lot of what we hope to accomplish will not be statistically measurable.*

Reducing students' anxiety about mathematics allowed students to progress more smoothly through many of the initial stages of mathematics learning. Implementing comfortable environments for learning mathematics enabled students to make more-natural progressions to higher levels of mathematics.

Meeting the goal of having students take more mathematics courses during high school took several different forms. Since one goal of the mathematics program was to have all students learn algebra, East Collins removed all general- and business-mathematics courses from its curriculum. As a result, more first-year algebra classes were added, and students who would not have studied algebra under the former system enrolled in algebra classes. Additionally, the teachers proposed eliminating the honors courses and concentrating on challenging all students mathematically. The honors courses were eliminated after the second year of the project. One teacher commented on this situation:

> *Something else I've noticed and talked about with some of the other teachers here: there are students in my algebra 2 class who five, six, seven, eight years ago would never have made it this far in mathematics—would never have made it this far, and yet they're equally competing and doing well. Being successful.*

Teachers also reported that girls seemed to be more engaged in mathematics classes. One teacher mentioned that she had read an article that cited research indicating that teachers pay more attention to boys. "They raise their hands all the time, they're usually the ones commenting." She noted that as part of the [mathematics] project, the teachers were trying to "avoid those traps," and work equally with all students. She added that "it's kind of interesting how it's turned around. The girls are doing a little bit better than the boys are."

The documenters gave the following summary of their visit:

> *The teachers at East Collins appeared to feel that their efforts to change students' attitudes had made some difference. They had evidence that more students were continuing in mathematics after first-year algebra and saw students as less anxious and as more responsible for their own learning. Most of the twenty-two students we interviewed seemed to support the teachers' feeling. Two students, in particular, exemplified the change in attitude. Both students were categorized by their teachers as average students, yet after taking a first-year algebra course taught with the PSML philosophy, both decided to take two mathematics classes—geometry and second-year algebra—during their junior year to earn an academic diploma. Previously, neither had thought about seeking an academic diploma.*

Motivation and Communication at Bedford Middle School

Bedford Middle School is one of twenty-four schools in a large urban district serving a school population that is 90 percent Hispanic and 9 percent Caucasian (1992–93 statistics). From the average of approximately 13 800 students who attend school daily in the district (90 percent of the total enrollment), about 9 100 (66 percent) qualify for free lunches and 1 400 (10 percent), for reduced fares. The district has fourteen elementary schools (grades K–5), four middle schools (grades 6–8), and two high schools (grades 9–12), plus a prekindergarten, two alternative secondary schools, and a center for handicapped students.

Reaching out to include more minority and underrepresented students in meaningful mathematics experiences may require that teachers upgrade and strengthen their own background in mathematical content and pedagogy. Such was the situation in the Mathematics Teaching Project (MTP) (see p. 7) when the district sought to develop an understanding of appropriate algebraic and geometric concepts among its elementary and middle school teachers. The documenters who visited Bedford Middle School reported that to achieve this goal, the project planned to assist teachers in these ways:

- Increasing their confidence to teach mathematics content
- Enhancing their strategies for teaching the essential elements
- Motivating them to try innovative techniques
- Emphasizing the importance of having high expectations of their students in mathematics

There were three stages of the MTP: (1) a preliminary study and discussion of the *Standards* documents, (2) a three-week summer MTP class, and (3) periodic follow-up meetings to share and discuss ideas that would further the professional development of the teachers and complement the implementation efforts.

Motivation to learn mathematics was an important theme in the reform at Bedford Middle School, and it was especially important to teachers and administrators that students discover that mathematics was fun. Teachers at Bedford Middle School were sold on the benefits of MTP and the way it was helping them motivate students in their classes. One particularly enthusiastic teacher explained,

> *Math should be fun.... Choose teachers who love math, choose teachers who love kids, and put them together.*
>
> —Algebra teacher
> Bedford Middle School

Even though Bedford Middle School was situated in one of the poorest areas of the district, there was a sense of pride in the educational community that teachers, administrators, and students had shared in creating. Documenters observed,

> *The double doors opened into the front hallway of an old, yet well-kept, two-story building. Murals, apparently painted by students, decorated the hallways; the floors were worn but clean; and there*

was no evidence of trash or graffiti anywhere. Although the office was buzzing with activity, the atmosphere was pleasant—certainly not at all frenetic. The school day was about to begin, but there were no hall monitors in evidence, nor did we hear ringing bells.

Teachers recognized the importance of communication in mathematics and believed that students were enjoying mathematics more and becoming more confident in their mathematical ability because so many avenues of communication were available. Small-group activity was one of the more important ways in which they shared their mathematics with one another. Teachers took pride in the way that students communicated mathematics to one another. One teacher volunteered the following during a classroom visit by documenters:

I walk around the room saying to myself "OK, Kate is really trying, she's not getting it, but she's really trying to communicate what she's doing." She's already passing right there because she's attempting to understand. And so she talks to her partner for a little bit, and her partner sets an idea off. Now the partner has just gotten an ace grade for understanding what Kate didn't understand and being able to communicate it to Kate. So now they both understand. Even if they don't get the answer right, as far as I'm concerned, they both succeeded that day. And they know that.

Teachers and administrators at Bedford Middle School faced many obstacles. For example, students' test scores on statewide testing ranked among the lowest in the state. As a result, much of the focus observed by documenters was on developing computational skills in mathematics. Still, there was a sense from the five MTP teachers in this school that they were on the right track toward motivating students to enjoy mathematics.

SUMMARY

Secada (1992) views the current system of mathematics instruction in the United States as operating to perpetuate and broaden the inequities that exist for the "poor and ethnic minorities." What is needed, according to Secada, is that reform efforts "should first become effective with these students, and then applied to other populations" (p. 654). Efforts at Pinewood High School and at Bedford Middle School are examples of reform efforts aimed at these populations of students.

The mathematics supervisor for Pinewood saw elitism in mathematics instruction as the single most important obstacle to mathematics reform, and the data indicated that reform efforts in that school were addressing the concerns of Secada (1992) (i.e., the emphasis on reform by piloting the Equity 2000 program to ensure that all students complete mathematics courses through algebra and geometry).

The efforts to provide equity in mathematics in two of the four schools previously mentioned does not seem aligned with Secada's position. The pattern at East Collins and Desert View was to first get the reform effort established for those students who were considered to be not at risk. Then, if successful with the "mainstream" students, the reform efforts would be extended to include those students considered at risk (i.e., ethnic minorities and the poor). Both high schools were making efforts to get more and more underrepresented students into more-demanding mathematics classes. At East Collins, general mathematics courses were disappearing by attrition as enrollments in algebra and geometry classes continued to grow.

Evidence shows that a major shift in pedagogy is offering more-interesting mathematical experiences to students in all four sites, which may be one of the principal factors in bringing about a more positive student attitude toward mathematics. It should be kept in mind that the observations of the documenters were intended to capture what members of the local communities perceived as being important as they went about the business of changing their mathematics programs. It was not the purpose of documenters to evaluate the degree of success that these schools were demonstrating in dealing with issues of equity.

Equity will continue to be an elusive part of the reform of mathematics education unless there is a constant effort at many levels. What evidence will be convincing that teachers are addressing issues of equity in mathematics classes? What are the characteristics of such mathematics classes? Will equitable provisions for students in mathematics necessarily include the development of an infrastructure as advocated by Weissglass (1996) or the transformation suggested by Noddings (1995)?

No ready-made formula is available for implementing programs for all students to learn mathematics. The collaborative process of reflection among educators is certainly indicated as one of the important ingredients of such reform (Johnson 1995; Knapp, Shields, and Turnbull 1995; Weissglass 1996). Figuring out a process, rather than a rule, must be determined by the local context in which the provision for equity is to take place.

Any significant change in which we deal with equity in mathematics education must necessarily begin with individual teachers. As suggested by Weissglass (1996), this change will require teachers to go within themselves and work to undo subtle—and not-so-subtle—attitudes and prejudices that have developed as a result of our various cultural settings. Teachers will need forums in which to share their changing attitudes about issues of equity. A high school teacher summed up the tenor of change:

> *I think this is all going to get better, too, as we all get some experience with the materials we have and the means that we have for different avenues—if we can find some time to talk to each other about it.*

The examples from the R^3M study provided snapshots of how individual schools and teachers were addressing issues of equity. All the reports showed early attempts at trying to solve a very complex problem. Without the kinds of interactions described by Weissglass (1996), we may never really understand the levels of prejudice and bias that continue to exist in our schools under the guise of conventions—conventions that prevent equitable opportunities in mathematics education.

Chapter 8

WHY IS CHANGE SO HARD?

Laura Coffin Koch

I taught for four years in another school, and then I went to the district office and worked with teachers, actually in the areas of reading and writing for four years and then left the district office because I felt like I needed to be back in the classroom.... I hadn't been anything but a traditional math teacher before I went out of the classroom. I had taught fifth grade and decided I wanted to teach young children. So that first year was a year of knowing that things should look differently, especially in mathematics. So mathematics has changed a lot for me in the last three years, and now I believe that children should construct their own understanding and that my role is to nudge them or help them see how connections can be made. Or maybe just sit back and watch them figure things out more than anything.... You heard me ask the kids today, "How is what you are doing mathematical?" and there's two reasons I ask that. The teacher part of me wants to make sure they're not fooling around and that they can justify what they're doing, but the other part of me knows that sometimes they say things about what they see of mathematics and they really can explain themselves and I wouldn't have ever come up with that. So I think the kids have helped me see that things are mathematical that I might not have seen before.

—Mathematics specialist
Parker Springs School District

A lot of time a good idea comes up and everyone gets excited and for the next two or three years it is there, and then, it just kind of fades away.

—Teacher
Parker Springs School District

Educational reform often begins with a vision. Sometimes it is a shared vision, and sometimes it is a vision held by a single person: a district administrator, a school principal, or a teacher. True reform is not like a book that can be discussed for its merits and then be put aside for a discussion of another topic or issue. Reform is not a one-time happening; it is an ongoing process. Reform takes on a life of its own. It requires change at all levels and in every facet of a school, a school district, an institution of higher education, a state, or a nation.

Nowhere is reform deeper, more personal, or more threatening than with teachers. The teachers are the ones who must use the new materials; understand the teaching implications; change how they interact with students, other teachers, and administrators; and defend their actions to all who are involved with the school system. As Sarason (1971, p. 193) states, "educational changes depend on what teachers do and think—it's as simple and complex as that." This chapter discusses the reform process by highlighting the types of challenges that teachers face as they attempt to change their teaching of mathematics. We draw on the experiences of the R^3M sites to gain insights about the challenges and tensions to be expected when sites use a "specialist" model for change, when central players have differences in viewpoint, when sites move beyond the beginnings, when mathematics is made a schoolwide priority, and when sites resolve curricular issues.

USING A "SPECIALIST" MODEL

Three years prior to the R^3M visit, the Parker Springs School District (PSSD) received a five-year, $500 000 foundation-funded grant for a districtwide grades K–3 mathematics specialist project. The intent of the project was to provide in-depth training and support for grades K–3 in-service teachers from each of the district's elementary schools. These specialists would then serve as agents of change to promote improvement in the district's mathematics education program. This investment had two fundamental guiding beliefs: (1) teachers who demonstrated a desire to improve their knowledge and skills would be able to create that desire in others, and (2) it would be possible to support these lead teachers in their change agency through information sharing and networking. Of course, even when teachers undergo profound changes in their own practice, they often find it difficult to effect change in others (Schifter and Fosnot 1993). Furthermore, teachers have few opportunities in their formal professional development to approach change agency systematically, drawing on a body of accessible skills and knowledge. Thus, this model for change is ambitious.

The special treatment and the relatively rich set of resources given the specialists as part of the program helped promote a sense of group responsibility. The specialists interviewed appeared to share the conviction that working collaboratively, they can support one another to reach a goal that is larger than what they might have achieved individually. The model seemed effective as a means of generating a new group of reformers within the district.

The relationship between the specialists and the other teachers proved to be complicated. The mathematics specialists were given special training and the resources to attend conferences and purchase classroom materials. Other teachers stated that they might be able to make changes if they were provided with the same types of support that the mathematics specialists received: resources, training, and manipulatives. One teacher stated, "It makes a big difference. It's hard to deal with a program when you don't have what it requires to do." One mathematics specialist told us that she was concerned about the number of visitors that came to observe her. She thought that other teachers in the school could not understand why she was singled out and "special."

Another tension existed between one specialist's perception of her role and the funded program's stated goals for the mathematics specialist. The specialist said that she believed her role to be that of a facilitator. She was willing to work with other teachers, observe their classes, and have them observe her, but she did not think that she was a better teacher. She just taught differently and was uncomfortable about forcing her ideas on anyone else. The goal of the program, according to the funding agency, was to effect widespread change within a school through the infusion of a single catalyst. The specialist was to be a mentor-teacher and an agent of mathematical change within a given school. Although the specialist's and the project's objectives appeared to be similar, they were not. The interpretation and implementation of those objectives created dissension.

How can a school deal with this kind of tension? What can a principal do?

When a teacher in a school becomes a leader, or an "expert" among peers, a shuffling of relationships and a sense of isolation may occur. The mathematics specialists at PSSD were not given any training or preparation to teach or serve as as mentors to other adults, especially their peers. It was not clear that the mathematics specialists were ready to do as Fullan and Miles (1992) suggest, that is, treat the other teachers in their schools as people who had opinions that mattered, ideas that counted, and ways to interpret and implement reform. For their part, the teachers had not been prepared for this introduction of new leadership. School-based discussions related to mathematics education might have supplied the teachers with information and opportunities for considering what impact the changes would have on them. Occasions for teachers to focus on their practice and on their own ideas about teaching mathematics prior to the specialists' work with them might have been useful.

How do teachers or administrators know if they are successful agents of change? How does a program prepare agents of change for when they may run into competing values and resistance from other teachers, school administrators, parents, and so on?

CHALLENGES OF CHANGE: STUDENTS' EXPECTATIONS

At East Collins High School, teachers faced challenges in dealing with students' expectations. Several teachers at the site mentioned that sometimes students have a difficult time initially accepting new roles that give them more responsibility for their own learning. One teacher reported that a student told her, "You're not teaching us." The teacher continued, "He was unhappy because I wasn't teaching him in the same old way that he expected. And it's just a whole different way.... I told him to hang in there." Another teacher added the following:

> *Some students were not used to the major techniques that are used with [the Perfect Situation for Mathematics Learning project].... They were used to having the teacher introduce the math technique, do some examples, work at their desk while the teacher walked around and helped them. What we did in [PSML] was just the reverse of that, they had to think on their own, reading first; and they ask, "Why am I reading?" They had to do it on their own. "But you haven't explained it to me yet," and then having to think through some things without that step-by-step direction. That's very frustrating for students, until they get used to that. They fight it.*

In this site, the process of sharing authority in the classroom was a learning experience for both the teachers and the students. But the teachers seemed committed to the process, and the students appeared to take hold of the autonomy they gained. One teacher summed up the thoughts of several teachers when she noted that what was "hard at first was being able to sit back and let the kids do instead of me do—having a lot more learner focus than teacher focus. But once I got used to it, it was more comfortable. And I've seen [the students] become more independent in their learning and a lot more confident with what they think they can do in math."

DIFFERING VALUES

Differences in viewpoint about what constitutes valuable mathematics can also generate various challenges. One specialist from the Parker Springs School District had been told by parents in the community that teachers in the grade level before hers did not recommend that their children be in her class because they would be "overly stimulated" in the hands-on environment. In their view children need to work on skills and use the textbooks. Several second- and third-grade teachers worried about consistency and the children's learning of the basic skills. The teachers also talked about how new mathematics programs would come and go and how some teachers were always willing to jump on the bandwagon without seeing any "proof" that the program would help children.

> *I guess I'm a very traditional teacher. I've been teaching a long time, and I've seen a lot of different programs come and go and I just find that in the last few years, when I think that it's good to be innovative and to offer the children a lot of different aspects and different ways of doing things, but we're finding a lot of basic skills are really slipping away.*
>
> —Second-grade teacher
> Parker Springs School District

This second-grade teacher's notion that students were not as well prepared as they once were was echoed by other teachers in the group. In contrast, the kindergarten and first-grade teachers, who emphasized a problem-solving approach, were pleased with how much their students knew and understood.

When teachers at different grade levels disagree about what should be emphasized, there will be problems with curricular continuity:

> *In second grade we have to start back doing things we used to do in first grade with children not knowing their basic mathematics until second grade, and if you want to do some of these problem-solving things, you have to know how to add and subtract to do those things.*
>
> —Teacher
> Parker Springs School District

In our visits we also detected differing beliefs about what mathematics was important for which students. At Garnett School, an ESL (English as a second language) teacher was using an interactive mathematics and science video program with her fifth-grade students. Because of a scheduling problem, she had to use videotapes of the live program, which prevented her students from calling in to ask questions of the presenters. She had hoped to have more students in her class, but another fifth-grade teacher who was supposed to provide students sent only a few. This teacher, who wasn't supportive of the reform initiatives at Garnett, decided that she would allow only those students who "wouldn't be hurt" by that "type of teaching" to participate.

Students' expectations and differences in viewpoint raise several important questions. Take a moment to reflect on the following questions before reading further. How do districts, schools, or projects resolve the lack of consistency from grade level to grade level (one teacher emphasizes the intent of the *Standards* documents, whereas the next stresses basic skills)? How do children become successful learners of mathematics when the emphasis of the program varies from year to year? How do teachers, schools, or districts know what they have taught and what students have learned? Are teachers able to build on the knowledge that the students have developed in previous grades, or do they believe that they must start from scratch? How do teachers justify to parents, administrators, and others that the students are actually learning?

BEGINNINGS: EXAMPLES FROM TWO CLASSROOMS

Barry Newton taught all the middle school mathematics—grades 6, 7, and 8—at Garnett School. He was trained as a music teacher but became recertified to teach middle school mathematics. Although he was considered to be the school's mathematics expert, he stated that he was still not comfortable with his own mathematics expertise. During the four days the observers spent in his seventh-grade mathematics class, anywhere from nine through fifteen students were present, usually equally divided between boys and girls. The class was the first hour of the day. On each day of observation, at least seven interruptions during the class period required his attention: students arrived late to school, announcements were made over the intercom, students needed to be excused from class for various reasons, and students needed to get supplies from their lockers in the hallway.

Mr. Newton provided stability and humor in the classroom. The behavior of the students was typical for students that age—fidgeting with hair, talking, appearing bored, and so on—but he kept them on task and tried to instill in them the importance of doing quality work. The classroom was set up with tables that were shared by three students. Locked cabinets covered two sides of the room, windows with bookshelves below covered the third side, and the front of the room had a chalkboard and a pull-down screen.

Mr. Newton was using materials from a university project in which he had recently participated. The materials consisted of sets of fraction bars and a worksheet with a series of questions related to finding and naming equivalent fractions. Each student worked with a partner, coloring in the correct color and size to match the fraction bars for the fractions equivalent to one whole. The process of coloring took most of the period. At the end of the class, Mr. Newton used an overhead projector and projector bars to lead an all-class discussion, but not enough time remained to complete the activity before the bell rang. He talked of his time-related frustrations:

> *I'm frustrated in that first of all, there is not enough time in a period to do everything that I would like to have accomplished on*

a day. And maybe it's partially my fault too because maybe I look to do more than I should, than in reality I'm going to be able to. Like today, I wanted to get much further in the fraction fact.... We started late and then one thing led to another and this always happens. So that maybe my expectation for doing the amount of work that we are going to do in a particular period is too big, but I think that there's just not enough time in the day to get through all that stuff that you need to and then the kids bring in a lot of baggage with them and might be ready to work and might not be ready to work. So you are torn whether you want to spend some time to help this kid out. This kid has a problem and is really upset about whatever reason. Whether you are going to spend part of your class period to do that or whether you are just going to write this kid off. So you are really torn because this kid might be one of your star students who's having a problem on this particular day. So what do you do? If you send this kid out or you don't pay attention to this kid, well then, the kid might not later on come back to you and talk to you as a friend.

The lessons that Mr. Newton used appeared to be grade-level appropriate, but the mathematics actually discussed in class was below what would normally be considered appropriate for grade 7. The observers were continually struck by the low level of mathematics that was taught, not only in this class but in most classes above kindergarten. The students we interviewed from Mr. Newton's class told us that they liked their mathematics class and knew that mathematics was important in their schooling, but they couldn't tell us why. Only one student had plans for continued education after high school.

Carina Olson taught fourth grade at Garnett School and seemed quite knowledgeable about the *Curriculum Standards.* She told us of the changes in her teaching and the workshops she had attended. She was very excited about the new directions in mathematics at Garnett and wished that other teachers were, too.

Ms. Olson stated that she believed that she needed to move slowly in her teaching of mathematics because the students were not grasping the material. Ms. Olson was teaching place value on the days she was observed. The students used base-ten blocks and a base-ten chart. All the students in the classroom were doing the same thing and were not allowed to explore or go beyond. Ms. Olson used the blocks to represent 29 on an overhead projector. She put two ten-blocks and nine units on the overhead projector and asked the students to do the same at their desks. She then went from student to student to be sure that each student had put the blocks exactly as she had. She then did three more examples: 28, 470, and 407. Once she had checked the students' charts, she had the students sit quietly and wait for the next number. When students started to play with the blocks or talked, she let them know that playing with the materials was inappropriate behavior. On the next day she varied the lesson only by putting a number, rather than the blocks, on the overhead projector. Ms. Olson explained that she wanted to be sure that the students knew the difference between 407 and 470. During the class, which lasted approximately fifty minutes, she gave the students four examples to follow, with the numbers ranging from 29 to 407. In the preobservation interview, Ms. Olson had explained the background of the lesson:

> *The day before they were very, very confused, and they were frustrated because they wanted to know and they didn't. Even 29, a number that is so easy. I would expect something different like 1000 or 2000, but 29? That they get so confused with that number is hard for me as a teacher.... I stuck to the Standards. I said maybe this would be better for them to see their hundreds, their tens, and their ones, and I did that worksheet. I went through the books and said maybe this is better so they can go over it again and repeat. And today is the same again—it's going to be a repetition, but in a different way.*

As teachers are beginning to try Standards-like approaches, management issues often become a challenge. How can this challenge be addressed?

Ms. Olson had been teaching for a long time, and change, she said, was not easy. She had a low opinion of the students' capacity for learning and of their future. She said that she tried to model the lesson for the students rather than just give them the book and tell them to do the problems. She believed that teachers should be aware that students use different approaches to doing problems. In the interview, Ms. Olson came across as a concerned and dedicated teacher with a sense of what should happen in a mathematics classroom.

MOVING BEYOND BEGINNINGS

Although the teachers of Garnett School spoke of the Standards, the ideas presented in the *Standards* document were not a prevailing influence in the actual teaching. Some teachers did stress problem solving and communication, but the types of communication we observed were at a low level. The teachers did ask the students many questions, but the questions usually required only short answers and little thinking or reasoning. The teachers were in the beginning stages of developing a pedagogical philosophy. Although several knew of the existence of the *Standards* and had a general sense of the *Standards,* they had a difficult time articulating their beliefs about teaching mathematics. This limitation is to be expected with teachers who are just beginning to make headway in reform.

MAKING MATHEMATICS A SCHOOLWIDE PRIORITY

One problem facing Garnett School was the lack of leadership and expertise in mathematics teaching. Ms. Evans, a fifth-grade ESL and mathematics teacher, served as the site liaison during the five-day R^3M visit. She was described as the driving force behind the reform movement at Garnett. Ms. Evans attempted to share what she learned in workshops with others—both teachers and administrative staff—in her school. Ms. Evans tried to encourage the teachers and staff by handing out a "problem of the week" for them to try in their mathematics classes. Although Ms. Evans felt committed to informing the teachers of what was going on and truly believed that changes needed to be made at Garnett, she had neither the expertise nor the additional time needed to make the change a schoolwide priority. No one else had been identified as a leader in Garnett's mathematics reform movement. The principal and the director of curriculum said that they did not know about the *Curriculum Standards* and were not comfortable with mathematics. The director of curriculum believed that it was important for change to occur in the teaching of mathematics, but she was not a mathematics person herself and could not provide the leadership that would have been be needed to move forward the reform movement and sustain the changes that were beginning.

The efforts of Mr. Newton and Ms. Olson, described in a previous section, were representative of the kind of teaching happening at Garnett: genuine caring about students and their learning, the ability to talk about what should happen in the classroom, the selection of appropriate resources and materials, the thoughtful preparation of mathematics teaching, and the willingness to reflect on their teaching. Several of the teachers talked with the R^3M documenters about wanting a mentor who would observe their classes and help them with the changes they were trying to make. They also talked about the necessity of schoolwide reform rather than just a few teachers working on their own. Many of the teachers stated that they were uncomfortable with the teaching of mathematics because of their own limited mathematics background and course work in mathematics. A few had had no mathematics in college and only the bare minimum in high school. Yet Garnett, as a school, had some level of identity as a place concerned with improving mathematics teaching and learning.

These teachers, who are at the beginning stages of change, recognized a need for an ongoing district or school-level program that would support teachers who were facing the need to grow professionally as they worked toward change in their mathematics program. How can such aspirations be supported and nurtured at the school level? What does it take to make mathematics a schoolwide priority?

RESOLVING CURRICULAR ISSUES

In the Parker Springs Schools District, teachers were aware of the district's core curriculum and all had used it in some way or another. From interviews, we learned that teachers who were in greater need of structured guidelines tended to follow a textbook rather than the district's core curriculum. The core curriculum seemed more useful to teachers who were interested in trying to approach the teaching of mathematics in different ways, and they used it as a tool to guide their innovations. Some of the teachers had started to use checklists based on the goals and objectives in the district curriculum. Several problems arose when teachers began to rely on checklists. First, the teaching of different areas of mathematics had become integrated for some of the teachers, who then found it difficult to "follow the strands" that were presented in the core curriculum. Second, a danger was present that teachers might teach only to the checklist, which could lead to minimal interpretation. How are curriculum guides and frameworks used at your site?

Garnett School used the district's mathematics curriculum that had been published in 1982: a list of skills and test items. At the time of our visit, none of the administrators seemed to know when it would be updated. The district kindergarten teachers had worked together and had formed a team of mutual support. They rewrote the kindergarten objectives to align with the NCTM *Standards* document. Do teachers generally feel empowered to recast district guides and curricula? What are the advantages and disadvantages?

SUMMARY

The tensions and challenges of change come in many forms. Some are personal issues, such as those faced by the mathematics specialist who felt a conflict between her teaching and the culture of the school. The specialist learned that what she thinks of her own teaching is important and that she does not always need to rely on others for validation. The tensions and challenges faced by the teachers at Garnett School involved working in a large district with little money to support its teachers. But the drive and the motivation of the teachers were strong and demonstrated that even with minimal support, uniform change can occur. Accountability and assessment are continuous and universal causes of tension. As schools and teachers transform their mathematics programs, they will need to be leaders in the development and selection of mathematical assessments that support their efforts.

It is not unusual for teachers to struggle in their attempts to try something new. The National Center for Research in Mathematical Sciences Education (1995, pp. 1–2) agrees:

> Like most organizations, schools and mathematics departments tend to be conservative in that they reward conventional practice. Those teachers whose practices depart from the conventional may be seen as threatening and may receive pressure from others to conform or to leave. Unless schools and departments change along with the teachers who are trying to change their individual practices, the large-scale long-term implementation of reform will encounter serious obstacles.

How will leaders resolve the tension that is sure to arise when principals, other school administrators, and other teachers are not supportive or do not share the vision held by some? If change is not always districtwide or schoolwide, how will the district work with other schools or jurisdictions toward the coherent improvement of mathematics education? In particular, how well developed must the organization plan be so that students moving between schools will not suffer because of the lack of a shared vision? What are the tensions and challenges faced by your district or school? What support is needed to help minimize these tensions and challenges?

Chapter 9

WHAT HAVE WE LEARNED FOR THE FUTURE?

Joan Ferrini-Mundy

We tried to get away from an arithmetic curriculum and get to a mathematics curriculum.

—Elementary school
mathematics specialist
Oldburg Elementary School

The R³M project provided a rare opportunity to look inside school districts, school buildings, and individual classrooms to learn what we could about the process of making change in mathematics education. The NCTM *Standards* documents gave us a useful set of guiding issues to focus our observations. The sites, in sharing their diverse efforts and experiences with Standards-based mathematics education, enabled us to collect a massive data set from which we might draw some conclusions, many questions, and some directions for the future. In this closing chapter, I highlight lessons from the R³M sites that might help other sites—individuals, groups of teachers, a school, a district—think about the following:

- Getting started in Standards-based mathematics education
- What to expect along the way
- Keeping it going

Clearly, no formula for this sort of implementation exists. Making change in mathematics education at the classroom, school, or district level is a very challenging and consuming process. What the R³M sites all share is a decision by someone or some group of people that a concerted effort to improve mathematics education was worth undertaking. The R³M study has given us snapshots that show how the sites got started and describe what happened as a result. We have very little

---

This chapter draws heavily on R³M project reports and the previous eight chapters for examples. Some of the quotations occur in previous chapters but may have been used to illustrate different issues.

conclusive data about the ultimate impact of these efforts on student learning as measured by "Standards-like" assessment tools; the study took place too early in the reforms for this to be possible. However, all of these seventeen sites passed the screening process for inclusion in this study because they could furnish some qualitative evidence that some aspects of mathematics teaching and learning were improved and effective in their setting. Possibly even more important as a first step, the teachers, administrators, and parents in the site were, generally speaking, pleased with the mathematics programs. If we might draw one conclusion, it would be that the strongest efforts we saw were those that recognized the challenge and worked to move forward consciously despite the difficulty. So within each of the following sections, we offer an array of possibilities and some questions to push you to deeper discussion.

GETTING STARTED IN STANDARDS-BASED MATHEMATICS EDUCATION

In our sites, the efforts that we saw at improvement and reform in mathematics teaching and learning had a number of interesting origins. The following examples seem characteristic of how general notions of a need for attention to mathematics moved to more-concrete events that helped in turn to launch site-level attention to mathematics teaching and learning. The influence of a single outside consultant, a carefully designed professional development program, an effort to obtain funding, or a beneficial partnership can be crucial.

Outside Consultants

> *Three years ago, a child in a small district in a the northeastern United States had severe learning difficulties in mathematics. The parents took the child to a consultant who specialized in learning-disabled children. After working with the child, the consultant, Dr. Madison Smith, agreed to write directions to the child's teacher, describing to them how the child learned mathematics most effectively. The teachers were impressed with the consultant's knowledge and asked him if he would come to a district meeting and give a presentation. Mr. Graham, the superintendent of the district, attended the consultant's presentation. Mr. Graham then asked Dr. Smith to look at the district curriculum, observe classes, note teachers' techniques, and see what suggestions Dr. Smith could make to improve the curriculum and teaching strategies.... Then the district hired Dr. Smith to train teachers and help the teachers design a new curriculum. Dr. Smith suggested that the teachers read the NCTM* Curriculum and Evaluation Standards, *which they did.*
>
> —R^{3}M documenter

We saw a number of sites that had been strongly influenced by an individual, usually from outside the site, who began to work with teachers in relatively small and circumscribed areas and whose sphere of influence broadened over time. In most of the R^3M examples of this type, the interaction with the outsider began over some fairly specific issue that was generated at the site and that was of particular interest to teachers. In other examples in the field, we know of teachers who as individuals forged relationships with mathematics educators outside their schools, often through participation in special institutes or other professional activities. These individual teachers often experienced very productive growth that came from their relationships with outside mathematics education leaders. The situation became especially interesting when they were able to bring that productive exchange into the school site and share it more broadly with a group of teachers.

Are individual teachers in your school already working with an outside expert or advisor on something that might be of interest to a broader group of teachers in your site? Could this relationship be broadened so that other teachers could benefit from the expertise? What pressing problems and questions do the teachers in your school have about their mathematics teaching?

Professional Development

> *Organized by the district's mathematics supervisors, together with a university professor, the course was designed to present the* Standards *to practitioners. The ultimate course goal was that teachers from grades 4–8 understand how to help all students succeed in algebra by grade 9, the assumption being that success is more likely if students have had appropriate experiences throughout elementary and middle school.*
>
> —R^3M documenter

In a number of the R^3M sites, the first steps toward a concerted effort to revise mathematics teaching were taken by means of the professional development of teachers. In the Oldburg School District site, we found a very specific, content-oriented focus for the professional development of teachers. This focus, together with the obvious commitment to K–12 articulation, distinguished this particular approach from others that we saw. In this site, the teachers' enthusiasm about their mathematics teaching, their appreciation for these well-planned professional development opportunities, and their camaraderie obviously contributed to what some would call a learning community.

Does your school have an area of curriculum that would serve as a useful focus to bring together grades K–12 teachers? Should you start with grades K–8 or 6–12 if your site has little experience with this cross-level articulation? How could you arrange a coherent sequence of professional development activities that over time would bring together teachers around an area of interest to them? Where would you find support and resources? Could your staff development dollars be allocated in this way?

Building Partnerships

The superintendent suggested that we needed the mathematics advisory committee, professionals to advise us.... So I wrote to twelve area corporations, described the problem, and asked them to provide someone in the corporation with a math-related degree. All twelve agreed. I remember the first meeting; we all had anxiety. We wanted them to advise us. I had all this stuff ready to give them, like our goals and curriculum. They said, "We don't want to see that. Let us spend some time debating and arguing, and we will come up with what we think should be written.... That council lasted three years. We have used that model three more times in our district at the beginning of an issue, not at the end of an issue, and it is now a model we use in anticipation of change.

—Assistant superintendent
Oldburg School District

Four committees were formed—the Mathematics Curriculum Research Team, the Mathematics Curriculum Writing Committee, the Mathematics Text Selection Committee, and the Mathematics Technology Scout Team. These committees collaboratively carried out the initial stages of the curriculum development process. This process started twenty-nine months prior to the implementation target date.

—R^3M documenter

In some sites, the process of improving and reformulating the mathematics program was undertaken through a formal process of revisions of curricular frameworks and guides. These processes were typified by the involvement of committees that included community members and parents, as well as school personnel. Where the reforms proceeded from such formalized and explicit goals and visions, the issue of teacher and community buy-in seemed to be less problematic. In some senses, proceeding along these lines is consistent with the way in which the NCTM *Standards* documents were developed—through a group-consensus process that included a wide range of stakeholders.

Is your district on a cycle of curriculum review that would allow a strong Standards-based perspective to be incorporated into the district's curriculum framework? How would you propose to use the *Standards* documents in this process? What community representatives and organizations would need to be part of the process?

WHAT HAPPENS ALONG THE WAY?

Every site in the R^{3}M study related some of the challenges, unexpected barriers, and positive outcomes as part of their initiative to enhance their mathematics programs. Parts of these accounts bore great consistency across the sites. We heard repeatedly about the ways in which the efforts at change were affecting the behavior and beliefs of children, teachers, and parents. We heard also about tensions and worries, both from the people in the sites and from the documenters at their distanced perspectives. Teachers were very conscious of what they often saw as a tension between emphasizing basic skills and teaching mathematics in a meaningful way; documenters raised occasional concerns about the superficiality of the reforms and about what they saw as an emphasis on pedagogy over mathematics content. The following excerpts from the documenters' stories convey these various "things to expect" in the hope that others embarking on this type of journey might have some sense of what it will be like along the way.

Parents

How did parents react to all the changes? A first-grade teacher at Oldburg remembered:

> *It was awful; parents weren't willing to accept it because we were actually changing the way we taught mathematics. They were afraid it was new math that happened after the sixties; they were afraid it was a fad. They were worried, and you can understand those worries because we were taking their children; our investment in their children's mathematics education was being tampered with.*
>
> *The specialist also involved thirty-five parents in a course of five two-hour classes where parents worked on the same activities that their children did in school. One parent who took the class because she had no idea what her child was doing in mathematics is now a Chapter 1 tutor at Oldburg and uses what she learned in the class in her tutoring.*
>
> —R^{3}M documenter

> *Handbooks for parents were developed for the K–5 and 6–12 programs. These handbooks addressed the most often asked questions that parents had about the new mathematics curriculum.*

These handbooks also stressed the role that technology would play in the new mathematics curriculum.

—R³M documenter

As we saw in chapter 5, parents' role and involvement in the reforms are essential, and sites handled this area in a range of ways. Generally, it seemed that educating parents to understand and appreciate the nature and intent of Standards-based mathematics education was the most productive strategy.

Are there key parents in your school who should be engaged early on in the improvement efforts? What are the concerns about basic skills likely to be in your community? Do you see a way to begin your reform that is likely to be appealing and acceptable to parents rather than be threatening and controversial?

Teachers

The teachers, like the students, seem relatively conservative, and cautious about change. They worry that the students will not get enough practice on basic skills, and they are concerned about saying anything that might offend the school administrators.

—R³M documenter

Those teachers who were active participants in the MTP claim they endorse the Standards and a problem-solving orientation to teaching, but they are reluctant to fully implement these ideas in their classrooms for fear that student scores on the state-mandated test will not be good.

—R³M documenter

We saw among teachers a wide range of belief, confidence, and commitment to the tenets of Standards-based reforms. The issue of basic skills was a central one for virtually every site. Teachers struggled mightily to enact some of the ideas and practices they believed promising while remaining very concerned that they might be taking time away from students' learning of important basic skills. Sites coped with this dilemma in various ways, but where this issue became an explicit topic of discussion and attention, teachers seemed to find ways to balance their efforts at reform with their commitment to their students' basic learning.

Why do people believe that the *Standards* documents advocate against basic skills? How do children acquire basic arithmetic skills? What approaches do teachers find most successful in helping their students know basic skills and facts?

It's hard to change. It's much easier to stay with what you know; you feel comfortable with that. So I think take it easy on yourself and do one thing at a time and then feel like you've accomplished that and then move something else in all the time. Get that area so that you feel good about it, next year add another area, then another area.

—Teacher
Deep Brook Elementary School

We met teachers who were unsure of their commitment to these directions but who were at least willing to try them. In one of the recent reports of the Third International Mathematics and Science Study (TIMSS), Schmidt and others (1996) define the notion of "characteristic pedagogical flow"—a deeply entrenched, cultural style of teaching that is pretty much standard in mathematics classrooms in the United States. This "characteristic pedagogical flow" involves what most of us might have experienced ourselves as mathematics students—introduction of the lesson, some group practice, and then time to practice individually. The TIMSS researchers found that this style prevails in mathematics classrooms in the United States. The shifts called for in the *Standards*—both mathematical and pedagogical—place an enormous demand on teachers to deviate from something that is deeply entrenched and supported by most aspects of the educational system. This demand on teachers should not be underestimated.

How can teachers support one another as they struggle with change? Where can teachers look for help in answering the questions that parents are asking? Is the notion of taking on one focused area of your teaching at a time a reasonable one? Where would you start?

Children

The students seem to want to know more. They have more questions about things. I find that children don't want the answers given to them any more. They are much more active now because they are involved in every lesson.... They're doing, they're asking, and they're getting involved. So I see a happier, more curious, and more problem-solving-oriented student.

—Teacher
Deep Brook Elementary School

Documenters came away from most sites with a very clear picture of students who seemed pleased with their mathematics classes. Repeatedly we heard that students liked "this way" of learning mathematics, that they found it interesting and fun to work with their peers, and that they enjoyed the activities and no longer found mathematics boring. Perhaps these attitudes mean that they found mathematics more meaningful. Of course, helping students feel motivated and getting them to enjoy mathematics are only part of reforming mathematics education; the ultimate goal is improved student understanding and learning of mathematics. A first step, for many teachers, did seem to be their observation that students were happier and more engaged—and teachers felt gratified and pleased with that development. Moving even further, to a point at which students can articulate how their understandings and skills are improving, will prove even more motivational for teachers who are looking to their students for evidence that these Standards-like directions are fruitful.

Is it good enough for students to think mathematics is fun? How can you tell, from student comments and from student work, what might be qualitatively new or different about their learning in these kinds of situations? How can you explain to parents and observers that Standards-based reform is more than play and activity?

OTHER ISSUES

> *Although the students are working in small groups, the tasks that the groups address are often very similar to those worked on individually before the reform effort started.... The vision so far is quite traditional—an emphasis on procedures with a contemporary overlay of calculator-based technology.... Although the methods used by the teachers were innovative, the content in their classrooms was generally not so.... In many of the classrooms we observed, children could be seen busily engaged with the computers in the room, playing with math mazes. However, given time to observe the children, it became evident that the exercises they were engaged in were low level and that the purpose of the task was lost. Many of the children were engaged in "drill and kill" activities that had little relevance to their math learning.*
>
> —R³M documenter

A number of the documenters' reports and comments raised concerns about what they called superficial efforts at reform—what some chose to describe as "talking the talk" only. The examples mentioned here were similar to a number of others that we heard: sites' using technology extensively, but for drill-and-practice exercises; teachers' using manipulatives with enthusiasm, but sometimes incorrectly and in rote ways; teachers' abandoning textbooks because they interpret the *Standards* documents to say that they should. Similar examples arose in all sites. Note that for every such comment, generally two or three other very specific examples of lessons *did* suggest that meaningful, high-quality mathematics teaching might be under way.

These instances represent, in my view, quite natural examples of teachers' or sites' entering the reforms through carefully and explicitly chosen pedagogical avenues using technology, manipulatives, discussion, and small groups. Given the newness of these approaches for many teachers, along with a relatively small amount of curriculum material explicitly designed to capitalize on these pedagogies and teachers' lack of prior preparation, my view is that these observations are not cause for alarm. What might be of concern would be the teachers' failure to push beyond these initial efforts and their lack of the skills and knowledge to reflect sufficiently on the nature of their practice to judge whether they are moving in directions they truly wish to go.

What does it mean to help students gain "mathematical power"? Are manipulatives, small groups, and technology alone necessarily the means for doing so? How can teachers decide if their pedagogical innovations are actually helping students gain mathematical power? How might the content recommendations made in the *Standards* documents become as salient for teachers as the pedagogical recommendations?

> *A few teachers in the upper grades do continue to use textbooks, but the teachers in the lower grades have opted to discontinue the widespread use of textbooks.... Although the Deep Brook School teachers agreed that this type of program requires more planning time because of the lack of textbooks, they remain no less passionate about their mathematics program. They realize they must supply the students with problems, projects, and related activities. However, they feel that this lack of dependency on textbooks forces them to become more collaborative with each other and more creative in their teaching.*
>
> —R^{3}M documenter

> *When we talk to people, it's very hard to separate the two [the ... texts and PSML]. And we know, [the other cochair] and I know and the other teachers sort of know, and [the assistant principal] knows that people looking at the program from outside, it's hard for them to separate the two, because they're.... The materials were exactly what we wanted. It amazed me pleasantly that a committee had listed all the things that we thought the program should have and what we were looking for in the book and the community support of the parents and all that. And then [a teacher] brought in these books, and it was almost like the books were written based on what we had described. It just fit very well.*
>
> —Cochairs, mathematics department
> East Collins High School

In several of our sites, especially at the elementary level, we found that one interpretation of the Standards-based reforms made frequently by teachers is that textbooks should not be used and that teachers should select and adapt activities from various sources. In contrast, in several other sites, mostly secondary schools, teachers and administrators made conscious decisions that certain textbook series were "Standards-based" and, therefore, were good choices. In many instances, we saw teachers mix elements from textbooks and adapt and invent their own materials and activities.

Do the *Standards* documents actually make specific recommendations about the use or nonuse of textbooks? When teachers do extensive adaptation and invention of their own activities, how do they ensure mathematical coherence over the course of the school year? Are there ways to experiment with inventing and with replacing textbooks that will guarantee that nothing important is omitted?

KEEPING REFORM GOING

Although the efforts at mathematics reform played out differently in each site and were highly dependent on the context and people involved, a few striking similarities in features we saw across the sites seemed to be part of the process of reform and possibly crucial for sustaining attention to, and progress in, the efforts at mathematics education improvement. These features appeared sometimes to evolve naturally as part of the culture of the site and sometimes to be deliberately initiated or catalyzed. We saw enough instances of ongoing professional collaboration among teachers in the context of their daily work and examples of teachers "making space" for specific experimentation in their practice that I choose to highlight some examples.

> *Taking advantage of the school's unique weekly schedule, where classes meet every other day for ninety-minute periods, they established an elective mathematics class so that students who chose could study mathematics on a daily basis by enrolling in their regular mathematics class (every other day) and in an elective mathematics class (on the alternate days). The elective class, called Logic, provided an opportunity for the teachers to "teach all the neat and fun things that exist in math" [that] they never had the time to teach before.*
>
> —R^{3}M documenter

> *Within the structure of the school day, teachers of the same grade level are provided with a block of common planning time at least three times per week to design specific lessons as a group, create student projects, share uses of manipulatives and technology,*

and collaborate on themes and activities to be used in all their classrooms.

—R³M documenter

We also saw many examples of sites finding creative ways to "make space" for teachers to experiment with the new pedagogies and with new mathematics content. We saw a very inventive mathematics-English class for sophomores at Scottsville. We saw teachers at Desert View find ways to add projects to their standard curriculum. What these examples have in common is the opportunity for teachers to work on changes in their practice in a sort of protected environment, often right alongside their standard teaching. The interesting result, in several instances, was that teachers were able in these "miniexperiments" to sometimes see startling positive effects on student learning. They would sometimes see, for example, that students whom they did not expect to understand mathematics could do well; they would also see that students could work with others. The teachers seemed to find the confidence to try some of their newly learned approaches in their "regular" teaching.

Is there any way to provide space for teachers to explore Standards-based teaching outside of their regular classes? An after-school enrichment program? A mathematics club? Team teaching with a teacher who is doing something that others might want to learn? Introduce a new course elective for students? Build more time into the mathematics class somehow? Take over a study period?

You can't have a one-shot staff development activity and say, "Now everybody's trained, go out and do it." It's got to be day in and day out, and the only way that you can do that is through collegiality in which staff members help staff members and that's got to be done on a daily basis. So I hope that after Dr. Smith, we will continue to focus on models that allow time for staff to work together and time for staff to teacher together as well.

—High school principal
Queensborough School District

Elementary school teachers also have a common grade-level lunchtime. One elementary teacher explained that during lunch hour they often talk about curriculum issues. They also have Thursday afternoons where they meet, as she explains, by grade levels for at least an hour.

—R³M documenter

You definitely need to spend at least a half an hour a day reflecting on what those children did, where you want to get them, looking at those resources you have and maybe grabbing a neighbor and saying, "OK, now what are you doing in math this coming week?" I've never really spent too much time thinking about what I'm doing in mathematics. It was too easy before. Now we're actually scratching our heads saying, "OK, now where can I really take this so it makes the biggest difference?"

—Teacher
Queensborough School District

But perhaps the most important support for the teachers comes from their work in "divisions"—especially the one for algebra teachers, and the division for geometry teachers, too. Division meetings provide an opportunity for teachers to share materials and teaching tips; the teachers also do some visiting in each others' classrooms. The change in atmosphere and the mutual support were noted by several teachers as an important impetus for sustaining the reform and for extending it to those teachers who have been reluctant to participate.

—R^3M documenter

In every site, "keeping it going" depended, to some extent or another, on the opportunity for, and commitment to, ongoing, substantive interactions among the teachers around the issues they were facing collectively in their efforts to introduce new content and pedagogy into their mathematics teaching repertoire. This is closely related to the notion of reflecting on one's practice, in consultation with peers. We heard of conferences taking place at the coffee machine, with teachers sharing their ideas and experiences. We observed team teaching, peer observations, sharing in the teachers' room—all activities that seemed both to grow out of, and to contribute to, a "collaborative community of practice."

What are the norms in your site for sharing professional discussions about mathematics teaching? How would you introduce these sorts of discussions? Which colleagues are likely to be interested in talking about their mathematics teaching? What might you use to get such conversations started?

The principal was complimented by the teachers for his dissemination efforts which included providing research and other professional writings in his interactions with them individually and collectively. He often achieved this by placing copies of articles and documents in their mailboxes on weekly basis.

—R^3M documenter

Several sites, in addition to having fairly formal professional development initiatives in place that supported the mathematics education program changes and in addition to the very informal but important sharing and reflection among teachers, indicated that other elements, such as "sharing material through the mailboxes," were contributing to an ongoing professional conversation about mathematics teaching and learning in the site. Repeatedly we heard how teachers were reading journal articles together, discussing them, and relating them to their own experience.

Could you start "sharing material through the mailboxes"? Could you share an interesting article that you have read? How would your colleagues react? Could you propose a discussion of an article over lunch? Could your principal arrange class coverage for a small group of teachers for one hour a month so that people could get together? What factors in your school would help support such activities? What factors would work to hinder them? How might the factors change over time? What other types of informal activities would be possible in your school or district?

CONCLUSION

In the initial discussion of a project to monitor the response to the NCTM *Standards* documents, we had a notion that we could find sites that were having success in their efforts at mathematics education reform and that by studying and describing those sites we could help to paint a more complete picture of the status of Standards-based reforms in the United States. The picture we have assembled is complex and difficult to summarize in any form. Clearly, work of this type must be continued. Ultimately, student achievement, as measured on both traditional and Standards-based measures, must be carefully linked to particular practices and curricular changes to even come close to "proving" that Standards-based reforms are successful. But if success can be measured also by evidence of truly committed teachers, parents, children, and administrators working daily in their settings to enact and interpret the *Standards* documents' visions of what is to be valued in mathematics education and to overcome the tensions, challenges, and barriers that can hold them back, then what we have seen in the R^3M project offers much hope for the impact of Standards-based reform.

What does the future hold for the mathematics program at your school? What advice might you have for others undertaking this process? What evidence is needed to "prove" that Standards-based reforms are successful? What support is needed to move forward?

REFERENCES

Apple, M. "Do the Standards Go Far Enough? Power, Policy, and Practice in Mathematics Education." *Journal of Research in Mathematics Education* 23 (1992): 412–31.

——. "Reflections and Deflections of Policy: The Case of Carol Turner." *Educational Evaluation and Policy Analysis* 12 (1990): 247–59.

Ball, D. (1992). "Implementing the NCTM Standards: Hopes and Hurdles." In *Telecommunications as a Tool for Educational Reform: Implementing the NCTM Standards,* edited by C. Firestone and C. Clark, pp. 33-49. Washington, D.C.: The Aspen Institute, 1992.

——. "Teacher Learning and the Mathematics Reforms: What We Think We Know and What We Need to Learn." *Phi Delta Kappan* 71 (1996): 500–58, 1996.

Cohen, D. "What Is the System in Systemic Reform?" *Educational Researcher* 24 (1995): 11–17, 31.

Corbett, D., and B. Wilson. "Make a Difference with, Not for, Students: A Plea to Researchers and Reformers." *Educational Researcher* 24 (1995): 12–17.

Ferrini-Mundy, J., and L. Johnson. "Standards in Mathematics: Issues in Documenting and Monitoring Progress." Durham, N.H.: University of New Hampshire, 1994.

Ferrini-Mundy, Joan, and Thomas Schram, eds. "Recognizing and Recording Reform in Mathematics Education: Issues and Implications." *Journal of Research in Mathematics Education* Monograph no. 8. Reston, Va.: National Council of Teachers of Mathematics, 1997.

Fullan, M. *The New Meaning of Educational Change.* 2nd ed. New York: Teachers College Press, 1991.

Fullan, M., and M. Miles. "Getting Reform Right: What Works and What Doesn't." *Phi Delta Kappan* 73 (1992): 745–52.

Haberman, M. "The Pedagogy of Poverty versus Good Teaching." *Phi Delta Kappan* 73 (1991): 290–94.

"High Standards for All Children and Assessments Aligned with These Standards." *Education Week,* 8 January 1998, 80–81.

Indiana Department of Education. *Mathematics Proficiency Guide.* Indianapolis: Indiana Department of Education, 1997.

Joftus, S., and I. Berman. *Great Expectations? Defining and Assessing Rigor in State Standards for Mathematics and English Language Arts.* Washington, D.C.: Council for Basic Education, 1998.

Johnson, L. "Extending the National Council of Teachers of Mathematics' 'Recognizing and Recording Reform in Mathematics Education' Documentation Project through Cross-Case Analyses." Ph.D. diss., University of New Hampshire, 1995.

Joyce, Bruce R., Richard H. Hersh, and Michael McKibben. *The Structure of School Improvement.* New York: Longman, 1983.

Knapp, M., P. Shields, and B. Turnbull. "Academic Challenge in High-Poverty Classrooms." *Phi Delta Kappan* 76 (1995): 770–76.

Ladson-Billings, G. "The Multicultural Mission: Unity." *Social Education* 56 (1992): 308–11.

———. "What We Can Learn from Multicultural Research." *Educational Leadership* 51 (1994): 22–26.

Lesh, R., and S. J. Lamon. "Trends, Goals, and Priorities in Mathematics Assessment." In *Assessment of Authentic Performance in School Mathematics,* edited by R. Lesh and S. J. Lamon, pp. 3–15. Washington, D.C.: American Association for the Advancement of Science, 1992.

Lesh, R., S. J. Lamon, F. Lester, and M. Behr. "Future Directions for Mathematics Assessment." In *Assessment of Authentic Performance in School Mathematics,* edited by R. Lesh and S. J. Lamon, pp. 379–425. Washington, D.C.: American Association for the Advancement of Science, 1992.

National Center for Research in Mathematical Sciences Education. "Reform at the School Level." *NCRMSE Research Review: The Teaching and Learning of Mathematics* 4 (1995): 1–2.

National Council for Accreditation of Teacher Education. *Introduction to NCATE's Standards.* Washington, D.C.: National Council for Accreditation of Teacher Education, 1994.

National Council of Teachers of Mathematics. *Assessment Standards for School Mathematics.* Reston, Va.: National Council of Teachers of Mathematics, 1995.

———. *Curriculum and Evaluation Standards for School Mathematics.* Reston, Va.: National Council of Teachers of Mathematics, 1989.

———. *Professional Standards for Teaching Mathematics.* Reston, Va.: National Council of Teachers of Mathematics, 1991.

———. *Principles and Standards for School Mathematics: Discussion Draft.* Reston, Va.: National Council of Teachers of Mathematics, 1998.

National Research Council. *Everybody Counts: A Report to the Nation on the Future of Mathematics Education.* Executive summary. Washington, D.C.: National Academy Press, 1989.

New Hampshire Department of Education. *K–12 Mathematics Curriculum Framework.* Concord, N.H.: New Hampshire Department of Education, 1995.

Noddings, Nell. "A Morally Defensible Mission for Schools in the Twenty-first Century." *Phi Delta Kappan* 76 (1995): 365–68.

Peak, Lois. *Pursuing Excellence.* U.S. Department of Education, National Center for Education Statistics no. 97–198. Washington, D.C.: U.S. Government Printing Office, 1996.

Ravitch, Diane. *National Standards in American Education: A Citizen's Guide.* Washington, D.C.: The Brookings Institute, 1995.

Robitaille, D., W. Schmidt, S. Raizen, C. McKnight, E. Britton, and C. Nicol. *Curriculum Frameworks for Mathematics and Science.* Third International Mathematics and Science Study Monograph no. 1. Vancouver, B.C.: Pacific Educational Press, 1993.

Romberg, T. "NCTM's Standards: A Rallying Flag for Mathematics Teachers." *Educational Leadership* 50 (1993): 36–41.

Rosenstein, Joseph, Janet H. Caldwell, and Warren D. Crown. *New Jersey Mathematics Curriculum Framework.* New Jersey Mathematics Coalition and New Jersey Department of Education, 1996.

Sarason, S. *The Culture of Schools and the Problem of Change.* Boston: Allyn & Bacon, 1971.

———. *The Culture of the School and the Problem of Change.* Boston: Allyn & Bacon, 1982.

Schifter, D., and C. Fosnot. *Reconstructing Mathematics Education.* New York: Teachers College Press, 1993.

Schmidt, W., et al. *Characterizing Pedagogical Flow.* Dordrecht, Netherlands: Kluwer Academic Publishers, 1996.

Schram, Thomas, and Loren Johnson. "Walking Together on Separate Paths: Mathematics Reform at Desert View." In *The Recognizing and Recording Reform in Mathematics Education Project: Insights, Issues, and Implications. Journal for Research in Mathematics Education* Monograph no. 8, edited by Joan Ferrini-Mundy and Thomas Schram, pp. 49–70. Reston, Va.: National Council of Teachers of Mathematics, 1997.

Secada, W. "Race, Ethnicity, Social Class, Language, and Achievement in Mathematics." In *Handbook of Research on Mathematics Teaching and Learning,* edited by Douglas A. Grouws, pp. 623–60. New York: Macmillan Publishing Co., 1992.

Shanker, Alfred. "Where We Stand." Internet communication dated 9 February 1997.

U.S. National Research Center. *Third International Mathematics and Science Study.* Report no. 7. Washington, D.C.: U.S. National Research Center, 1996.

U.S. Office of Education. *Pursuing Excellence: Initial Findings of the Third International Mathematics and Science Study.* Washington, D.C.: U.S. Office of Education, National Center for Educational Research, 1996.

Weiss, I. *A Profile of Science and Mathematics Education in the United States, 1993.* Chapel Hill, N.C.: Horizon Research, 1994.

Weissglass, J. "No Compromise on Equity in Mathematics Education: Developing an Infrastructure." Unpublished paper. Santa Barbara, Calif.: University of California at Santa Barbara, 1996a.

———. *Ripples of Hope: Building Relationships for Educational Change.* Santa Barbara, Calif.: Center for Educational Change in Mathematics and Science, University of California at Santa Barbara, 1996b.

R^3M PROJECT DOCUMENTERS

(Affiliations at time of project involvement)

Joan Ferrini-Mundy, Project Director
University of New Hampshire

Clem Annice
University of Canberra
Australian Capital Territory, Australia

Gabrielle Brunner
Milton Academy
Milton, Massachusetts

Mark Driscoll
Education Development Center
Newton, Massachusetts

Beverly J. Ferrucci
Keene State College
Keene, New Hampshire

Julie Fisher
National Council of Teachers of Mathematics

Karen Graham
University of New Hampshire

Loren Johnson
University of California at Santa Barbara

Laura Coffin Koch
University of Minnesota

Diana V. Lambdin
Indiana University

Linda Levine
Orange County Public Schools
Orlando, Florida

Anita Long
University of New Hampshire

Joanna O. Masingila
Syracuse University
Syracuse, New York

Douglas McLeod
San Diego State University

Geoffrey Mills
Southern Oregon State College

Lois Moseley
Region IV Educational Service Center
Houston, Texas

Thomas Schram
University of New Hampshire

Paola Sztajn
Indiana University

Patricia P. Tinto
Syracuse University
Syracuse, New York

Julian Weissglass
University of California, Santa Barbara

Bonnie Whitley
William Byrd High School
Vinton, Virginia

Terry Wood
Purdue University—West Lafayette
West Lafayette, Indiana

R³M ADVISORY BOARD

(Affiliations at time of project involvement)

Mary M. Lindquist, Chair
President, National Council of Teachers of Mathematics

Deborah Ball
Michigan State University

Joan Ferrini-Mundy
University of New Hampshire

James D. Gates
Executive Director, National Council of Teachers of Mathematics

Marilyn Hala
National Council of Teachers of Mathematics

Linda Levine
Orange County Public Schools
Orlando, Florida

Edward Silver
University of Pittsburgh

Donald M. Stewart
The College Board

Lynn A. Steen
Mathematical Sciences Education Board

Robert Witte
Exxon Education Foundation